Rosângela Lopes Borges
Ronaldo R. V. de Oliveira

Environmental perception in higher education

Rosângela Lopes Borges
Ronaldo R. V. de Oliveira

Environmental perception in higher education

Case study in Biological Sciences (2013-2016)

Imprint

Any brand names and product names mentioned in this book are subject to trademark, brand or patent protection and are trademarks or registered trademarks of their respective holders. The use of brand names, product names, common names, trade names, product descriptions etc. even without a particular marking in this work is in no way to be construed to mean that such names may be regarded as unrestricted in respect of trademark and brand protection legislation and could thus be used by anyone.

Cover image: www.ingimage.com

This book is a translation from the original published under ISBN 978-613-9-72482-6.

Publisher:
Sciencia Scripts
is a trademark of
Dodo Books Indian Ocean Ltd. and OmniScriptum S.R.L publishing group

120 High Road, East Finchley, London, N2 9ED, United Kingdom
Str. Armeneasca 28/1, office 1, Chisinau MD-2012, Republic of Moldova, Europe
Printed at: see last page
ISBN: 978-620-7-92744-9

INDICE

INTRODUCTION

There's a lot of talk about the environment, diversity, biodiversity, sustainability, preservation, environmental perception and environmental education. However, it is difficult to really define what the "environment" is. Authors such as Geraldino (2014) define it in three different areas: The environment of things, The environment of the living and The environment of humans.

The breadth of the definition of "environment" can be seen in the authors' definition of it as a place where there is no life (stone), where there is life (bacteria/animals/plants) and where human beings exist (symbolic thought). It can be seen that the environment involves all possible places on and off the planet.

Man projects himself onto the environment and the environment projects itself onto man. There is a relationship of complicity between the two. The dependence of human beings on the environment is notorious. From the 18th and 19th centuries onwards, the environmental impacts caused by man grew rapidly. That's when it was realised that humanity's survival would depend on nature and the care given to it.

A wave of environmental preservation movements was triggered, and since then human beings have become more and more concerned about where they live and the consequences of their impact on nature. Nowadays, we are advised to avoid wasting natural resources, to economise, to reuse, to recycle, to use public transport more, to avoid disposable items and to reduce wasteful consumption.

As a result, society's responsibility towards the environment began to spread. Environmental responsibility is a set of attitudes, whether individual or collective (companies), aimed at the sustainable development of the planet. In other words, these attitudes must take into account socio-economic growth adjusted to environmental protection, guaranteeing sustainability and the preservation of natural resources for future generations.

For a long time it was thought that natural resources would never run out and that nature would withstand any human intervention. However, it is now recognised that naturally, many of the Earth's resources can be regenerated or recycled, but this only happens over time - usually a very long time and one that man does not have at his disposal. Natural resources are contained within the boundaries of the planet and are therefore limited.

This cycle of destruction could be balanced, but by a sufficiently intelligent human species with a distinctive ecological perception. The term perception refers to the way in which each individual feels the environment around them, valuing it to a greater or lesser extent. Given this diversity of thought, it can be seen that perception is inherent to each human being, as they react and respond differently due to their values, which are influenced by culture, history, religion, social class and a host of others.

Environmental perception is man's becoming aware of his surroundings so that, by perceiving them, he learns to protect and care for them in the best possible way. This perception is also directly influenced by the environment. With this in mind, Environmental Education was created.

Environmental education is the process through which individuals build social values, knowledge, skills, attitudes and competences aimed at conserving the environment, which is a good for the common use of the people and essential to a healthy quality of life and its sustainability. There is no better place to offer this than in schools.

Environmental Education is supported by law in the school environment, where documents such as the PCNs and the Transversal Themes address environmental issues that seek to raise awareness and change the behaviour of students and the community in general. This teaching is aimed at all levels of education, from nursery school to higher education.

Universities are expected to be able to promote, through Environmental Education as a "new" field of knowledge, criticism of the current way of appropriating and transforming nature in the construction of the civilising process of modern life. That these HEIs stimulate questioning in the social and natural environment, thus implying epistemological and pedagogical possibilities for teaching, seeking effective solutions to socio-environmental problems.

To this end, higher education institutions must promote diversified teaching practices in order to arouse students' interest in the subject and, consequently, change their behaviour and spread a different "culture". It will be up to teachers to seek continuing training to enable them to direct their students' scientific and technological research, producing new knowledge and opening up new horizons.

According to scholars such as Araújo, Oliveira and Nogueira (2005); Krasilchik (2008) and Dornfeld (2015), undergraduate programmes in Biological Sciences have a dual responsibility in relation to Environmental Education. The first refers to the teaching of Environmental Education offered in other courses, seeking to raise awareness, change perceptions and behaviour. The second is the responsibility of training future teachers who will disseminate this new model of life to their students. It is understood that the latter is more difficult to achieve.

In view of the concepts raised in this work on environmental perception, the following problem was raised: What was the students' environmental perception like when they started the Biological Sciences course and what is it like when they finish? Has the course taught them and changed their attitudes and behaviour in relation to the environment?

It was hypothesised that when students start the course, their environmental perception is individualised and limited. It is assumed that during the course this notion of the environment and sustainability will be broadened. It is also assumed that the teachers do not have the necessary knowledge and training for environmental education.

This work is justified by the educational, social, economic, cultural and political importance of Environmental Education within Higher Education Institutions. It is understood that studying students' perceptions of the environment in which they live makes it possible to determine whether learning has been significant and whether teaching practices have achieved the expected results. The study is therefore organised as follows:

Chapter I - Protocol Framework: a description of the problem, the research questions, the justification for the research, the delimitation and limitation of the topic, as well as the research objectives.

Chapter II - Theoretical Framework: discusses the definition of perception; types of perception; environmental perception; environmental perception in schools and in higher education; environmental education in the world, in Brazil, in schools and in higher education; interdisciplinarity in environmental education and pedagogical practices for environmental education.

Chapter III - Methodological Framework: the type of research, the approach adopted, delimitation, population and sampling, inclusion criteria, hypothesis, data collection techniques and instruments, sources of information, ethical considerations and how the data collected will be processed.

Chapter IV - Analytical Framework: here the data collected during the research will be analysed and discussed. They are: the analysis of the interview applied to the teachers, as well as the analyses of the questionnaires applied to the students in 2013 and 2016.

Chapter V - Conclusion: the aim of this section is to conclude the work with the final considerations and recommendations that could be made after applying and analysing the field research, in the light of the theoretical and bibliographical framework.

CHAPTER 1

PROBLEM STATEMENT

1.1 DESCRIPTION OF THE PROBLEM

Perception is understood as the act, effect or faculty of perceiving, acquiring knowledge of something through the senses, understanding and listening. Environmental perception encompasses understanding the interrelationships between the environment and individuals, i.e. how society perceives its surroundings, expressing its opinions, expectations and proposing lines of behaviour (OKAMOTO, 1996).

Transversal Themes (2001) states that as humanity increases its capacity to intervene in nature to satisfy growing needs and desires, tensions and conflicts arise over the use of space and resources. Nowadays, there is also the aggravating factor of rapid technological advances that make overconsumption and the accumulation of rubbish possible.

In all areas, the environment, nature and its environmental resources have become a priority for governments' political and economic planning. Society therefore imposes rules on the growth, exploitation, distribution and conservation of these resources in order to reduce environmental impacts (BRASIL, 2001).

The National Curriculum Parameters (PCNs) define sustainability as "[...] the use of renewable resources in a qualitatively adequate manner and in quantities compatible with their capacity for renewal, in viable economic solutions [...] adequate quality of life for all". (BRASIL, 1997, p. 178).

All the recommendations, decisions and international treaties on environmental issues highlight the importance attached by leaders around the world to Environmental Education as a means of creating and applying sustainable ways of life. Working with this theme in the classroom aims to contribute to the formation of a citizen who is aware of and capable of dealing with the socio-environmental reality (DIAS, 2000).

School institutions must therefore provide a healthy educational environment that is coherent with what they want their students to learn. It is also up to these school environments to guarantee situations in which students can put their ability to act into practice. To this end, teachers need to be trained and qualified to deal with these issues in the classroom (FORTUNATO; SHIGUNOV NETO, 2015).

We therefore realise the importance of Environmental Education for the academic environment, specifically for Biological Sciences students who will be future Science and Biology teachers in primary schools.

This raises the following question: What was the students' environmental perception like when they started the Biological Sciences course and what is it like when they finish? Has the course taught them how to change their attitudes and behaviour towards the environment?

1.2 RESEARCH QUESTIONS

The formulation of the research questions establishes the hypotheses and the bridge between the problem and the design, data collection methods and analyses (SOARES, 2003).

1.2.1 Generic questions

What is the environmental perception of students on the Biological Sciences course, has this course broadened their perception and promoted changes in the way they perceive the environment in which they live?

1.2.   2Secondary Questions

How did the students perceive the environment when they entered the Faculty?

How do students perceive the environment at the end of the course?

Has the Biological Sciences course offered more than individual sensory perception?

What knowledge do teachers have about environmental education?

Have the interdisciplinary practices used by teachers been enough to change student behaviour?

1.3 RESEARCH OBJECTIVES

The objective of the study in a research project is a declarative statement that specifies the direction of the research according to the level of knowledge established in the field in question. It specifies the key variables, the target population and the direction of the research (SOARES, 2003).

1.3. 1General Objective

To characterise the perception that Biological Sciences students have of the environment and the results of the Environmental Education offered by teachers.

1.1.1 Specific objectives

To assess the environmental perception of the students and whether the Biological Sciences course has promoted progress in their individual perception;

To investigate the evolution of the students' environmental perception after joining the course;

To analyse the teachers' knowledge of their students' environmental education.

To investigate the pedagogical practices used by teachers when teaching their classes.

1.4 BACKGROUND

This study is justified in the sense that Environmental Education, as prescribed by the PCNs (1997), is of the utmost importance for the future inhabitants of the planet to be able to perceive the environment and intervene in its preservation and quality of life.

The Transversal Themes on the Environment (2001, p. 189) argues that "Access to new information makes it possible to rethink practice". It is therefore possible to note the importance of socio-environmental approaches in the academic sphere, as well as the ongoing training of teachers.

In this way, it is understood that assessing the perception of students on the Biological Sciences course will help to evaluate teachers' approaches to environmental issues and, consequently, it will be possible to see whether the teaching offered at this Higher Education Institution has been significant and whether it has helped to broaden the environmental perspective of its students.

1.5 DELIMITATION AND LIMITATION

The aim was to carry out a survey of Biological Sciences undergraduates in order to assess the evolution of their environmental perception.

The instruments used to develop this study were obtained specifically from the Biological Sciences class that started in 2013 and finished in 2016. We also took into account whether they had already studied Environmental Education (EE) in the 7th term and Environmental Planning (EP) in the 8th term.

In this context, there is free access to Higher Education Institutions as this is where the author of this study works. It is understood that his position as a lecturer in the Biological Sciences programme may facilitate access to the data to be investigated.

CHAPTER 2

THEORETICAL FRAMEWORK

2.1 CONCEPTUAL FRAMEWORK

Perception: A feminine noun originating from the Latin perceptione and describing the act, effect or ability to perceive something. Environmental perception is man's awareness, so that by perceiving the environment in which he is inserted, he learns to protect and care for it in the best possible way.

Environmental Education: The process of education responsible for training individuals who are concerned about environmental problems and who seek the conservation and preservation of natural resources and sustainability.

Biological Sciences course: A higher education course that covers content specific to the Biological Sciences, such as biodiversity, molecular biology, genetics, among others. It takes the form of a Bachelor's or Licentiate's degree.

2.2 HISTORIC MILESTONE

For hundreds of years, human beings have exploited natural resources unbridled in the face of an economy geared towards overconsumption. As a result, we are facing one of the biggest problems in the history of the planet: the scarcity of natural resources and climate change (DIAS, 2000).

The concern that something should be done about environmental degradation only began between the 1960s and 1970s. At that time, the high levels of pollution and environmental degradation in various parts of the world due to industrialisation began to be analysed (OKAMOTO, 1996).

Since 1986, with the introduction of CONAMA Resolution 001 in Brazil, many projects with the potential to impact the environment have been obliged to prepare an Environmental Impact Study (EIS) and its respective Environmental Impact Report (RIMA) as part of the licensing process for their implementation and operation.

Brazil's environmental legislation, developed through the National Environmental Policy, gained more strength and consolidation with the advent of the new Constitution of the Federative Republic of Brazil in 1988, which dedicated a special chapter to environmental issues and encompassed all the legislation in force in the country.

Environmental issues also began to permeate the school environment, since today's students would be tomorrow's future residents and citizens. With this in mind, the National Curriculum Parameters (Parâmetros Curriculares Nacionais - PCNs) were created in 1997, addressing issues related to the environment and sustainability as an interdisciplinary component.

Currently, authors such as Fortunato and Shigunov Neto (2015, p. 24) argue that "Learning environmentally means fitting the environmental challenge into the very dynamics of learning and not as an occasional complement." Schools should consider nature as an integral part of life on the planet.

As such, the aforementioned authors suggest that the school should be designed in an environmentally correct way; that the contents of the curriculum should naturally have an environmental reference; that administrators, teachers and students should be environmentally configured; that there should be concern for the school's natural environment; that there should be permanent evaluation of the products and the surrounding environment; and that care should be taken with the school's surroundings.

2.3 PERCEPTION: SOME CONSIDERATIONS

This section will discuss the definition of the term perception, as well as environmental perception in its various spheres. It is understood that it is extremely important to know these concepts before discussing Environmental

Education specifically in the academic sphere.

2.3.1 Definition of perception

The term Perception comes from the Latin "perception" found in the Portuguese language dictionary Ferreira (2014) as: "which has the purpose of perceiving the act", i.e. the ability to know independently of the senses; ideas; sensations, among others.

It finds a wide range of meanings from these dictionary definitions. These concepts range from the perception of stimuli to intuition, idea and image. Both are distinct categories in philosophical discourse. These professionals constantly reflect on the history of human thought, the expression of which gives rise to theories such as idealism, empiricism, realism and materialism (MARIN, 2008).

Hochberg (1973, p. 11) apud Marin (2008, p. 206) "perception is one of the oldest subjects of speculation and research in the study of man".

According to the author, perception is an attempt to explain an individual observation of the world. The study of perception predates the science of psychology, and its first scholars were physiologists and physicists.

In the field of psychology, specifically behaviourist psychology, the use of the senses in the recognition of an object and response to stimuli was studied. The focus of studies at that time was on the physical and biological mechanisms that led people to a certain behaviour or choice (HOCHBERG, 1973).

Behaviourism focused on entirely observable behaviour as a response to environmental stimuli. Environmental psychology, in its initial phase, adopted the principle of biophysical and behaviourist studies. Both are behaviourist approaches that gave rise to the first studies on perception. In an attempt to overcome the traditional tendencies regarding the relationship between human beings and their environment, some scholars have criticised this approach:

Environmental psychology is distinguishable from the traditional fields of perception because most of the stimuli it uses, although they come from the physical environment, have been extracted from it and applied to captive individuals in a controlled laboratory situation [...] This is a very appropriate procedure for scientists, and it teaches us a lot about the characteristics of stimuli and the various ways in which people respond to them, but it tells us very little about the environment (LEE, 1977, p. 13).

According to Lee (1977), there should be a distinction between environmental psychology and other studies of perception. For him, studying a situation in isolation resulted in nothing, and said nothing about the environment, since situations were controlled in the laboratory by artificial and unnatural stimuli.

Marin (2008) cites Gestalt[1] , in Environmental Psychology, which according to her was very significant for the study of environmental psychology. The Basic Laws of Gestalt are: Similarity, Proximity, Continuity, Pregnancy, Closure and Unity. The last two are exemplified for a better understanding of the study of Environmental Psychology (Figure 1, next page).

According to the authors, in the Basic Law "Law of Closure", elements that appear the same are interpreted as a complete object, even though they are only line segments. In the "Law of Unity or Unification" *the* human mind tends to instinctively fill in the empty spaces of abstract images so that they are understood by the human mind (PERLS; RALPH; GOODMAN, 1997).

Figure 1 - Basic Gestalt laws

Source: (PERLS; RALPH; GOODMAN, 1997)

During the 19th century, environmental psychology studied the variation

---

[1] Gestaltism or the Psychology of Form is a doctrine of psychology based on the idea of understanding the whole in order to perceive the parts. Gestalt is a word of Germanic origin, roughly translated as "form" or "figure".

of behaviour in built environments such as rooms, houses, buildings, cities, etc. Psychological maps were even drawn up, which were discussed taking into account the conditions of stress and satisfaction in residential and/or natural environments. Scholars understood that the same person could act differently when faced with the same stimulus if they were in a different environment.

In the 20th century, there was a theoretical consolidation of environmental perception studies, which spread to other areas of knowledge. Del Rio and Oliveira (1999) state that Geography, Urban Planning and Architecture recognised the importance of environmental psychology and began to apply it to space. Other areas of study such as Education, Sociology, Engineering and Forest Resources also began to study it.

At the beginning of the 1970s, research into environmental psychology spread internationally with the creation of the UNESCO group[2] Man *and Biosphere* - 13 (MAB). According to Melo (2005), this project emphasised the study of the perception of the environment as

the management of places and landscapes of importance to humanity. The MAB has studied the relationship between populations and the environment in various cities in 40 countries, including the Brazilian city of Porto Alegre-RS.

Oliveira (2001) explains that in the 1970s, space, place and landscape began to be approached from a human perspective, as the concept of landscape was broadened. The idea of "landscape" was no longer restricted to nature, but involved human knowledge, awareness and affectivity. This theory comes from Humanistic Geography. In architecture, according to the architect Okamoto (2002, p. 9), the need arose to "create man's perceptive environment for his proper development, using an environmental language".

Fróis (2001) explains that according to the theories of Norberg-Schulz (1975), spatial awareness is based on people's experiences with the things and schemes around them. It stems from affective needs in relation to their

---

surroundings and is culturally determined. Space in the perception of an urban dweller is different from that of a rural dweller.

Perception is therefore understood to be the process of organising and interpreting the sensory data received in order to develop awareness of oneself and the environment. In educational studies, perception is part of the process of forming knowledge and, consequently, values (WHITE, 1994).

2.3.2 Types of perception

According to Monteiro and Ferreira (2009), perception is the brain function that allows individuals to organise and interpret their sensory impressions. In this process, meaning is attributed to the environment that surrounds it, based on a history of experiences. Revista Cérebro (2010, p.12) adds that "The process of perception begins with attention (a process of selective observation), as it is this that will capture certain external stimuli.".

This cognitive activity involves acquiring, interpreting, selecting and organising the information obtained, which may or may not depend on information from memory. Initially, the receptor organs detect the stimulus (a process known as sensation). Subsequently, this stimulus is translated into nerve impulses that are conducted to the brain, where they are processed. This takes into account factors both external and internal to the body (REVISTA CÉREBRO, 2010).

Monteiro and Ferreira (2009) explain that perception of a given subject changes as new information is acquired. According to the authors, even in the absence of a change in the stimulus (influence from third parties), perception can change due to the subject's experiences. It is in the brain that sensory information is processed and given meaning.

However, Brain Magazine (2010) warns of two situations if the brain or sensory organs are not in total balance with the characteristics of the environment. The first concerns "a deficient perception" in which what is perceived does not actually correspond to the stimulus offered. The second is non-existent perception, which occurs when the individual is unable to perceive anything at all in relation to the stimulus.

Perception should not be confused with memory. Memory is based on past events, while perception is entirely related to episodes in the present. The same goes for thought and consciousness. Perception arises from a stimulus coming from outside, unlike consciousness and thought, which are more intimate and subjective processes that exist even without a stimulus, so it is only necessary to resort to memory (BRAGA, 1998).

It is possible to understand that the presence and condition of the individual are very important for the perception of the stimulus. For this reason, according to Monteiro and Ferreira (2009), there are different types of perception that allow you to adapt to the information received and thus organise representations of the world.

Brain Magazine (2010) defines the types of perception as follows:

a) Visual perception: perception of shapes, spatial relationships, colours, light intensity and movement.
b) Auditory Perception: based on analysing the perception of timbres, pitches and frequencies, the perception of sound intensity and volume.
c) Olfactory perception: the perception of odours through the nose is extremely important for our sense of taste, especially when eating.
d) Taste perception: perception of flavours by the tongue, usually associated with pleasure.
e) Tactile Perception: perception of objects and sensations through the skin, allows the presence, shape, size and temperature of objects in contact with the body to be recognised.
f) Temporal perception: perception of temporal durations, production of rhythms, temporal order and simultaneity. Often used in music.
g) Spatial perception: perception of distances between objects. It involves combining auditory, visual and temporal perception.

Perception is closely linked to consciousness. Consciousness is a function of the mind that allows the individual to study their surroundings. In order for it to come into being, it is necessary to use the five senses (sight, hearing, smell, touch and taste) as a means of connecting external stimuli. Cognitive functions arise from this process (OLIVEIRA, 2001).

2.3.3 Environmental Perception

There is another type of perception, in addition to those mentioned above, which is environmental perception. According to Coimbra apud Philippi-Jr et al (2004), it is the way in which each individual feels about the environment around them, valuing it to a greater or lesser extent.

The same authors, cited above, also consider environmental perception to be an awareness-raising process on the part of man, so that, by perceiving the environment in which he is inserted, he learns to protect and care for it in the best possible way.

Perception is recognised as the effect of perceiving and acquiring knowledge from something through the senses. Environmental perception encompasses "how society perceives its surroundings, expressing its opinions and expectations" and through these, lines of behaviour are woven, maintaining an interrelationship between the environment and the individual (AMANTE, 2001 apud FRAZÃO; SILVA; CASTRO, 2010, p. 158).

Studies that use environmental perception aim to investigate how people see, interpret, live with and adapt to the reality of the environment in which they live, especially when it comes to unstable or socially and naturally vulnerable environments (OKAMOTO, 1996). It is worth emphasising that, in this study, the definition of environmental perception goes beyond individual sensory perception.

Fernandes and Pelissari (2003) add that environmental perception can be defined as man's awareness of the environment. For the authors, it is from this perception that the individual begins to take an interest in protecting and caring for the environment. Each individual perceives the environment differently, and there can be individual and collective perceptions.

For this reason, according to the aforementioned authors, the study of environmental perception "is of fundamental importance so that we can better understand the interrelationships between man and the environment, his expectations, desires, satisfactions and dissatisfactions, judgements and behaviours". (FERNANDES; PELISSARI, 2003, p. 13)

It was through UNESCO, in mid-1973, that research into environmental perception began. It was realised that the greatest difficulties in protecting natural environments lay in the existence of differences in perceptions of values and importance between individuals from different cultures and socio-economic groups (FERNANDES; REZENDE FILHO, 2010).

According to (2001), environmental perception can be used to assess the environmental degradation of a given region, as in the case of the Alto da Bacia do Limoeiro, Presidente Prudente-SP. The analysis of the data collected made it possible to highlight and interpret the process of degradation, highlighting the omission of the public bodies responsible for licensing and monitoring urbanisation. According to the author, education and environmental awareness are emerging as "weapons" in the defence of the natural environment, helping to bring people closer to nature. It guarantees a future with a better quality of life, since it awakens responsibility (individual and collective) and respect for the environment in which we live.

2.3.4 Environmental perception in schools

Cabral (2015) argues that the advance of industrialisation and exacerbated population growth in various parts of the world have aggravated environmental degradation. This poses a threat to people's quality of life, making it necessary and urgent to find solutions to the problems of the natural and modified environment.

Environmental education emerged in the search for possible alternatives to environmental problems. Studies focused on this reality analyse the application of initiatives that awaken the concept of environmental education in various sectors, especially in school environments. However, the educational process must "[...] overcome the merely ecological vision, transposing the gaze to a more comprehensive dimension, with discussions of political, social, economic, cultural and environmental issues." (SPIRONELLO et al., 2012).

Dias (2000) recommends that an educational process should begin with a diagnosis of the local references and practices of the very people to whom the process will be directed. It should also involve broadening environmental cognition, so that people can understand, structure and learn about the subject.

The school, as an environment for transmitting and receiving knowledge, becomes the ideal space for developing knowledge, values, attitudes and attributes that are favourable to the environment. Environmental education is the fundamental tool for interacting in this process (SILVA; LYRA; ALMEIDA-CORTEZ, 2003).

2.3.5 Environmental Perception in Higher Education

Given the prominence of environmental issues in today's society and their close association with specifically scientific areas, especially the natural sciences, an emphasis on environmental issues has been realised and educational proposals for the sector have emerged (SCHMIDT et al., 2011). Environmental education and studies on environmental perception have been based on the document "United Nations Decade of Education for Sustainable Development[3] ".

The role of leadership is seen as important through the integration of teaching and practical learning on sustainability issues. Because sustainability is embedded in the economic, social and environmental context, its approach

---

[3] Document drawn up by UNESCO (2005-2014) The Decade of Education for Sustainable Development has at its core a simple idea with complex implications, because after living for centuries without worrying about the depletion of the planet's natural resources, we now have to learn to live sustainably. The great challenge is to encourage changes in behaviour and attitudes.

becomes interdisciplinary, as it emphasises the study of best practices in pursuit of sustainable development (UNESCO, 2005).

Education for Sustainable Development - ESD, according to with UNESCO (2005):

> [...] applies to all people, regardless of age. It takes place [...] throughout life, involving all possible learning spaces [...] from early childhood to adulthood. ESD requires a reorientation of educational approaches - curriculum and content, pedagogy and assessments. Learning spaces include non-formal education, community organisations and local civil society, the workplace, formal education, technical and professional training, teacher training, higher education, educational inspectors, political bodies [...] and much more (UNESCO, 2005, p. 20).

Furthermore, the same document warns that ESD should not be seen as "one more subject" to be added to an overloaded curriculum. It advises that it should be a global "whole school" plan, in which sustainable development is seen as a context for achieving the objectives. "Considering ESD as a red line that crosses the student's path throughout the education system - from pre-school to higher education - will maximise its impact." (UNESCO, 2005, p. 61).

Barbieri and Silva (2011) illustrate that the role of universities, colleges and other higher education institutions is of great importance in environmental education. Their participation is not restricted to discussion and raising awareness. However, it is transversally linked to showing the tools for practices that lead to changes in attitude towards sustainable development.

UNESCO (2005, p. 61) explains that "teacher training institutions [...] are often overwhelmed by changing and expanding curricula". Higher education teachers must not only be convinced of the need for ESD, but they must also have the methods to integrate ESD into their teaching practices.

The Global Higher Education Alliance for Sustainability (GHESP), with the support of the United Nations University, has begun to develop a pedagogical kit aimed at reorienting higher education towards sustainable development (UNESCO, 2005). Higher education must also take on a leadership role, putting into practice what teachers teach, integrated into teaching and learning. With academic research and investigation programmes, HEIs must function as places of research, since "Educational theory and innovative practices often emerge from academic research and investigation programmes." (UNESCO, 2005, p. 62). (UNESCO, 2005, p. 62).

Sibbel (2009), in a text published in the International Journal of Sustainability in Higher Education, says that the role of universities is to improve curricula so that they are able to offer students experiences capable of developing an awareness capable of associating their professional practice with their social, moral and environmental responsibility. These practices include planning and activities at higher levels, which will be more valuable if they are designed to support local and sub-national levels.

To better elucidate what these local and sub-national groups are, Table 1 below was drawn up using information from UNESCO (2005, p. 86-89):

Chart 1 - Local and sub-national groups involved in sustainable development.

Institutions and Organisations	Tasks and co-operation
LOCAL LEVEL	
Schools (universities), Associations, Co-operatives, Committees, etc.	Identify local challenges. Integrating local knowledge and practices. Promote the exchange of experience. Identify and implement learning strategies.
REGIONAL LEVEL	
Organisations, alliances, associations, universities, regional groups, etc.	Support the formulation of national policies. Promote private sector co-operation with other sectors. Learn and pass on common lessons from international experience and bring them to the regional reality.
NATIONAL LEVEL	
Government, Ministry of Education, Universities, NGOs, Private Companies and Associations, etc.	Promote a structure of regulations at national level. Budgeting and mobilising resources. Encourage public awareness. Linking national and local experience. Promoting awareness through the media.
INTERNATIONAL LEVEL	
Intergovernmental agencies, UNESCO, agencies, social networks and NGOs, etc.	Facilitate and participate in international and regional forums. Promote inter-regional learning and exchanges. Organising campaigns and publicity with international organisations. Setting standards and awarding indicators.

Source: Organised by the author (2017) adapted from UNESCO (2005).

It should be noted that universities offering higher education are part of local, regional and national groups. This means that changes can start at a local level, with seemingly insignificant actions that nevertheless change systems, and then spread throughout the rest of the country, having a more consistent and transformative effect, and can even reach an international level.

2.4 ENVIRONMENTAL EDUCATION

This section looks at the history of Environmental Education, covering its emergence in the world, in Brazil and in Higher Education. Understanding the emergence of this "science" is important for the application of the research.

2.4.1 The   history of environmental education

The first records of the use of the term "Environmental Education" date back to mid-1948, at a meeting of the International Union for Conservation of Nature (IUCN) in Paris. Environmental education really began to be defined at the Stockholm Conference in 1972, when it was put on the international agenda. Later, in 1975, in Belgrade (Yugoslavia), the principles and guidelines for the future were defined through the International Environmental Education Programme (HENRIQUES et al., 2007).

In 1977, the Intergovernmental Conference on Environmental Education was held in Tbilisi (Georgia). At this meeting, which was consolidated by Brazil, the definitions, objectives, principles and strategies for environmental education that have been adopted around the world to this day were decided (DIAS, 2000).

Dias (2000) explains that it was after the book "Silent Spring", released in 1963, that environmental education movements intensified. Journalist Rachel Carson, the author of this book, warned of the damaging effects of human actions on the environment, such as the loss of quality of life caused by the indiscriminate and excessive use of chemical products and their subsequent effects on the environment, such as the use of pesticides.

At the Global Forum in 1992, the Treaty on Environmental Education for

Sustainable Societies and Global Responsibility was drawn up, establishing a link between environmental education (EE) public policies and sustainability. It set out principles and an action plan for environmental educators. It emphasised participatory processes aimed at recovering, conserving and improving the environment and quality of life through collective and interdisciplinary work (HENRIQUES et al., 2007).

According to Reigota (2007), this treaty is very important because it was drawn up by civil society and because it recognises environmental education as a dynamic political process, in permanent construction, guided by values based on social transformation. He also adds Agenda 21[4] , approved during Rio 92[5] , as one of the guiding principles for environmental education, since this participatory planning process involves the country, state, municipality, region and sector, as well as planning for the future in a sustainable way.

Henriques et al. (2007, p. 13) recognise that "It has been recognised that, five years after Rio 92, the development of environmental education has been insufficient". It was necessary to reformulate the basic premises that guide education in the pursuit of sustainability. In 2002, the United Nations implemented the document "The Decade of Education for Sustainable Development (2005-2014)", which leverages existing educational policies, programmes and actions, as well as multiplying innovative opportunities.

2.4.2 Environmental Education in Brazil

Environmental education in Brazil began before it was institutionalised by the government. It was born in the mid-1970s through the isolated actions of teachers, students and schools. According to Henriques et al. (2007, p. 13), it

---

[4] A document signed on 14 June 1992 in Rio de Janeiro by 179 countries as a result of the "United Nations Conference on Environment and Development" - Rio 92, it can be defined as a "participatory planning instrument aimed at sustainable development".

[5] Conference on the planet's environment: sustainable development of countries. The United Nations Conference on Environment and Development (UNCED), held in Rio de Janeiro in June 1992, marked a turning point in the way humanity views its relationship with the planet.

"manifested itself through small actions by civil society organisations, city halls and state governments, with educational activities aimed at actions to recover, conserve and improve the environment".

In 1973, with the creation of the Special Secretariat for the Environment (Sema), the process of institutionalising Environmental Education in the Brazilian federal government began. In 1981, the National Environmental Policy (PNMA) established the need to include environmental education at all levels of education. This reinforced item VI of article 225 of the Federal Constitution (1988), which established the need to "promote environmental education at all levels of education and raise public awareness for the preservation of the environment".

Under the influence of Rio 92, two bodies were created in the executive branch to deal exclusively with environmental education: the MEC's Environmental Education Working Group (1993) and the Environmental Education Division of the Brazilian Institute for the Environment and Renewable Natural Resources (Ibama), which represented a milestone in the institutionalisation of the country's environmental education policy (DIAS, 2000).

In 1994, the Ministry of the Environment (MMA) was created. During Rio 92, the Brazilian Charter for Environmental Education was written, which recognised Environmental Education as one of the most important instruments for enabling sustainability as a strategy for the survival of the planet and, consequently, for improving the quality of human life (GUNTZEL-RISSATO et al., 2013).

Ibama and the MMA encouraged the formation of State Interinstitutional Commissions for Environmental Education. In 1994, the Presidency of the Republic created the National Environmental Education Programme (PRONEA), which was implemented by the MEC's Environmental Education Coordination and the corresponding sectors of the MMA/Ibama, responsible for actions aimed at the education system and environmental management respectively (REIGOTA, 2007).

In 1995, Conama - the Temporary Technical Chamber for Environmental Education - was created within the National Environment Council. In 1996, the Environmental Education Working Group was created and a protocol of intentions was signed with MEC, aimed at technical and institutional co-operation in Environmental Education (DIAS, 2000).

In 1997, the National Curriculum Parameters (PCN) were established. This document is used to support schools in drawing up their educational projects. Due to the need to deal with some urgent social issues, of national scope, called Transversal Themes (environment, ethics, cultural plurality, sexual orientation, work and consumption) opening up to the relevant reality of schools. (HENRIQUES et al., 2007).

With the creation of the General Coordination of Environmental Education (CGEA) at MEC and the Directorate of Environmental Education (DEA) at MMA in 1999, Law No. 9.795 was passed, which provides for the National Environmental Education Policy (PNEA) and in 2002, Law No. 9.795/99 regulates the composition and competences of the PNEA Management Body, thus laying the foundations for its implementation (GUNTZEL-RISSATO et al., 2013).

In 2004, the third version of the National Environmental Education Programme (ProNEA) was submitted for public consultation. This document involved around 800 environmental educators from 22 of the country's federal units. That same year, environmental education began to be integrated into the areas of Diversity, Indigenous School Education and Education in the Countryside, thus giving environmental education greater visibility and emphasising its cross-cutting vocation (REIGOTA, 2007).

Also in 2004, through the joint work of the MEC and the MMA, Environmental Education became part of the Curriculum Guidelines for Secondary Education and the Distance Education modules in the Education of

Youth and Adults (EJA). That same year saw the start of a new Multiannual Plan, the PPA 2004-2007 (GUNTZEL-RISSATO et al., 2013).

It's important to note that environmental education doesn't only take place in formal education, but in all educational spaces outside the classroom, in the family, the community, the church, the street, and associations, through socio-environmental activities, courses, workshops, events, seminars, conferences, projects, programmes, campaigns and many other actions that take place outside the school (DIAS, 2000).

2.4.3 Environmental Education in Schools

The Tbilisi Conference, 1977, also pointed out that environmental education in schools should seek out its own spaces inside and outside the school as teaching resources. Dias (2000, p. 214) quotes part of this text "t) Use the environment itself as a teaching resource. [...] Field trips and visits by students are indispensable in environmental education

However, teachers need to know how to exploit existing spaces both in the school and in the community. Dias (2000, p. 214) argues that "u) EE requires that teachers receive training and learn to use new pedagogical approaches. v) Intensive research is needed.

Continuing training for teachers helps to improve their classroom practice. "Schools, from nursery schools to postgraduate courses, are privileged places for environmental education," and should encourage research and practice through sustainable actions (REIGOTA, 2007, p.40).

Environmental education was made compulsory by Law 9.795 of 27 April 1999. The Environmental Education Law, as it has come to be known, states in Art. 2º : "Environmental education is an essential and permanent component of national education, and must be present, in an articulated manner, at all levels and modalities of the educational process, in a formal and non-formal character."

Since then, Brazil has been making efforts through guidelines and public policies to promote and encourage Environmental Education in primary schools, especially since the second half of the 90s. According to the classification of the 2004 School Census, Environmental Education in Brazil is applied through three main modalities: Projects, Special Subjects and Insertion of Environmental Themes into Subjects (HENRIQUES et al., 2007).

However, it was through the National Curriculum Parameters - PCNs (1997) that Environmental Education was consolidated in schools. Environmental Education aims to be interdisciplinary and, at the same time, integrated into formal education. This integration into the education system brings us back to the Parameters which, in Article 22, ensure access to education and propose pedagogical and methodological principles to be observed by educational institutions."

One of the main objectives of the PCNs is that students, at the end of primary school, should be able to "perceive themselves as integral to the environment and as agents of its transformation, being able to contribute to its improvement". (PCN, 1999, p. 34).

The PCNs (1997) incorporated the Transversal Themes (2001) into conventional subjects, relating them to reality, transferring to the teacher the responsibility of educator with the aim of training the individual in citizenship. They comprise six areas: Ethics (Mutual Respect, Justice, Dialogue, Solidarity), Sexual Orientation (Body: Matrix of sexuality, gender relations, prevention of sexually transmitted diseases), Environment (The cycles of nature, society and the environment, environmental management and conservation), Health (self-care, collective life), Cultural Plurality (Cultural Plurality and the Life of Children in Brazil, the constitution of cultural plurality in Brazil, the Human Being as a
social agent and producer of culture, Cultural Plurality and Citizenship) and Work and Consumption (Labour Relations; Work, Consumption, Environment and Health; Consumption, Mass Media, Advertising and Sales; Human Rights, Citizenship) (BRASIL, 2001).

Pascarelli (2011) recounts the movement for "Ecopedagogy" that emerged at the "First International Meeting of the Earth Charter" from the perspective of education. Organised by the Paulo Freire Institute on 23 August 1999 in the city of São Paulo, it was approved by a civil society forum representing all peoples and is the document of planetary citizenship.

Ecopedagogy is a concept that is closely linked to sustainability. Through global thinking and a vision of the future, with proposals for preservation and appreciation combined with a positive attitude towards the planet (PASCARELI, 2011).

The Cross-Cutting Themes act as a unifying axis around which subjects are organised, and should be worked on in a coordinated way and not as a decontextualised subject in class. Their role at school is to facilitate, encourage and integrate actions in a contextualised way, through interdisciplinarity and transversality, seeking not to fragment knowledge into rigid blocks, so that education really is the means for social transformation (DIAS, 2000).

Gadotti (2000, p. 236) states that "curriculum content has to be meaningful to the student and will only be meaningful to them if this content is also meaningful to the health of the planet [...]". He also mentions the importance of working in the classroom on children's doubts, conflicting ideas and curiosities, trying to integrate school with life, offering solutions to problems, trying to harmonise students with their environment.

Berne (2004, p. 51) states that: "[...] Environmental education can and should be present in all subjects [...]. Finding ways to do this is a challenge for everyone." With this in mind and concern, the 5th Environmental Education Forum was held in November 2004 and its theme was teacher training. Ongoing teacher training is seen as the best way to successfully tackle the environmental issue in Brazil and around the world.

2.4.4 Environmental Education in Higher Education

Dornfeld (2015) says that nobody denies the importance of socio-environmental issues and environmental education. However, despite being widely publicised, degradation has increased considerably and there has been no materialisation of this reported and necessary cultural change.

This is no different in public and private higher education institutions. Valuable initiatives can be found in isolation, but they are small and solitary. The article "Environmentalisation of higher education institutions: environmental education contributing to the construction of sustainable societies" (SORRENTINO, BIOSOLI, 2014) considers some paths to be taken, called "transversal axes":

a) THE FIRST TRANSVERSAL AXIS: enabling environmentalisation is dialogue on utopias and values. Institutions cannot neglect this responsibility to provide each of their participants with an in-depth questioning of the consumerism that distances us from ourselves and to present a wide range of utopias of all times, encouraging each person to build their own values and virtues.

b) THE SECOND TRANSVERSAL AXIS: for building this environmentalisation of institutions is methodological. The United Nations Educational, Scientific and Cultural Organisation (UNESCO) suggestion can be adopted as a starting point, setting out four essential skills for the future professional student-citizen of all areas and regions: Learning to learn, Learning to do, Learning to be together and Learning to be.

c) THE THIRD TRANSVERSAL AXIS: the definition of a priority territory for the institution's activities, in which it fosters an educating collective made up of various other institutions and people who co-operatively formulate and implement a Political and Pedagogical Project that emulates demands for its Research, Teaching, Management and Extension activities.

Since its mission is professional education and the training of educators, the HEI plays a fundamental role in supporting the process of incorporating environmental education into other levels of education, through initial and continuing training, as well as extension and postgraduate programmes. "Environmental Education, in university curricula and practices, has a strategic meaning in the environmentalisation of education and society." (BRASIL, 2007, p. 25).

Higher education institutions are spaces for knowledge, research and teaching and, as a consequence, for transforming society through intellectual development and freedom of thought. Therefore, there is no doubt that HEIs must set an example in sustainability in the social, economic and environmental spheres (SOBRAL, 2014).

Dornfeld (2015) illustrates that:

> Several universities around the world have made efforts to promote sustainability actions on their campuses. In this sense, as a way of diagnosing the university's current situation, monitoring progress over time, assessing the effectiveness of sustainability programmes and reporting on the results of these initiatives, it is essential to define criteria and indicators, as these act as tools for assessing sustainability (DORNFELD, 2015, p. 05).

The research proposal, carried out in 2007, entitled "Mapping Environmental Education in Brazilian Higher Education Institutions:

elements for public policies" arose from the convergence of demands to draw up both guidelines for implementing the National Environmental Education Policy (PNEA) and strategies for consolidating Environmental Education within the scope of higher education. This research was divided into three main topics: Topic 1: institutional organisation and guidelines; Topic 2: formalised teaching, research, extension and management activities - courses, projects and disciplines; Topic 3. Diagnosis of facilitating elements and difficulties in implementing environmental education programmes (BRASIL, 2007).

Still on the subject of the research "Mapping Environmental Education in Brazilian Higher Education Institutions: elements for public policies", Henriques et al. (2007) summarise the main difficulties and facilities reported by the interviewees:

> • Main difficulties: i) resistance of various kinds, attributed **in part to the "immaturity" of EE as a discipline 9 which does not yet have a** consolidated theoretical and methodological framework; ii) lack of financial resources and

The same document discusses the need for public policies on Environmental Education for Higher Education to be articulated with other policies. In addition, it is hoped that "[...] education, science and technology policies will recognise environmental education as an interdisciplinary area of knowledge production that is scientifically valid and socially relevant, considering it as an area to be supported and stimulated." (BRASIL, 2007, p. 26).

2.4.5 Interdisciplinarity in Environmental Education

Environmental Education, as a proposal for the different levels of education, cannot run the risk of being treated in isolation as part of a subject (ARAÚJO; OLIVEIRA; NOGUEIRA, 2005). Nor can it be separated from life; however, it must permeate the curriculum in a comprehensive manner, and preferably be worked on in an interdisciplinary and transdisciplinary way.

The disciplinary sciences emerged at the end of the last century out of the need to respond to the fragmentation caused by a positivist epistemology. They came to the fore in 1912 with the founding of the Jean Jacques Rousseau Institute in Geneva by Edward Claparède, Piaget's teacher. A whole discussion was held about the relationship between the mother sciences and the sciences applied to education, and related notions emerged such as multidisciplinarity, intradisciplinarity, pluridisciplinarity and transdisciplinarity) (ARAÚJO; OLIVEIRA; NOGUEIRA, 2005).

Multidisciplinarity characterises traditional education, where content is organised independently, i.e. there is no relationship between the content and the different disciplines. Intradisciplinarity is understood in the educational sciences as the internal relationship between the "mother" discipline and the "applied" discipline. (ZABALA, 2002).

Pluridisciplinarity is a term that refers to the relationship between more or less related disciplines. Interdisciplinarity, on the other hand, can be understood as the relationship between two or more disciplines. Dias (200) characterises transdisciplinarity as more ambitious than interdisciplinarity. He says: "[...] for the time being, in fact, we should focus on

concern ourselves with interdisciplinarity, and then move on to transdisciplinarity" (DIAS, 2000, p. 25).

Nicolescu (1999) distinguishes three degrees of interdisciplinarity:

> a)a degree of application (e.g. nuclear physics methods transferred to medicine lead to the emergence of new cancer treatments;
> b) an epistemological degree (e.g. the transfer of methods from formal logic to the field of law produces analyses in the epistemology of law;
> c) a degree of generation of new disciplines (e.g. biophysics, biochemistry, psychopedagogy, etc). (NICOLESCU, 1999, p. 52).

The term interdisciplinarity in education no longer poses a problem because, when dealing with the same object of science, one science of education "complements" another. The same can be said for multidisciplinarity. It is the nature of the educational act itself, i.e. its complexity, that demands a multidisciplinary explanation and understanding (ARAÚJO; OLIVEIRA; NOGUEIRA, 2005).

According to the aforementioned authors, pedagogical action through "inter- transdisciplinarity" points to the construction of a school institution that is participatory and decisive in the formation of the social subject. The aim of the institution, despite environmental education, would be to articulate knowledge, experience, school-community, the environment, etc.

In this context, the Curriculum Guidelines for Secondary Education (OCEM), published in 2006, arose from the challenges and understanding of Biology as an essential subject for the basic education of every citizen. The recommendation is that the document should be the subject of clarification, dissemination and discussion in schools about the proposed content and practices (DORNFELD, 2015).

Among the strategies suggested by the OCEM (2006) to guide teachers in working with Environmental Education in a more enjoyable way are: environmental studies, games, seminars, debates, simulation and project development. The latter has been pointed out by educators and researchers in the field of environmental education as an efficient strategy that has been seen in some schools, but only occasionally (GUIMARÃES, 2005).

2.4.6 Pedagogical practices for environmental education

Environmental studies are strategies aimed at getting students out of the classroom and into the environment, which can be the school's surroundings, their neighbourhood, etc. This allows them to: assess the environmental conditions of the place studied, interview residents, draw up proposals to improve the conditions found and identify public administration bodies to forward the proposals and/or claims drawn up (OCEM, 2006).

According to the PCN (2002, p. 56), "Using games as a pedagogical tool is not restricted to working with ready-made games, in which the rules and procedures are already determined; but, above all, to stimulating creation by the students Games and play are motivating in the teaching-learning process because they are pleasurable activities that can get everyone involved.

Seminars and debates help students learn to express themselves and defend their ideas coherently, and also to respect the opinions of others. Both strategies involve research that can be carried out individually or in groups. There is a wide range of topics related to Environmental Education that can be debated in the classroom (SATO, 2003).

In the simulation, students can be guided by the teacher to propose and carry out simple experiments to test hypotheses about the content being worked on (OCEM, 2006). Students can carry out research and turn it into graphs or data, presenting it to the class, the school and even the community. Simulation is a teaching strategy that allows people to experience a representation of a real event in order to practise, learn, evaluate or understand these situations (PAZIN; SCARPELINI, 2007).

Classes with demonstrations" aim to transpose the cold limits currently set for formal, descriptive and axiomatic teaching towards a new scenario, rich in stimuli and strongly interactive, capable of reaching every student. It is believed that this type of lesson is a departure from traditionalism, in that it makes the content interesting and enjoyable, as it instils the student's curiosity right from the start (SAAD, 2005).

For Hernández (1998), the function of a project is to encourage the creation of strategies for organising school knowledge. "A project should not be a task determined by the teacher, but chosen and discussed by everyone, teacher and students" (OCEM, 2006, p. 27).

An inter- and transdisciplinary education project should be characterised by a general vision of education in a progressive and liberating sense. However, scholars such as Tristão (2002) have found that so-called interdisciplinary projects are no more than multidisciplinary. Starting from the students' interests is a "kick-start" to avoid making this mistake.

Another point to be emphasised that can guarantee the success of classes dealing with environmental issues is the care teachers take to arouse students' interest; they need to be motivated. From this perspective, Rossini (2003) emphasises that students' motivation depends on the satisfaction of their internal motives, but it also depends on external motivation.

In pedagogical practice, technologies and other means are used to mediate this pedagogical practice. In this search, the possibility of doing pedagogical practice at two extremes using material/resources is visible. Some of the resources used in the teaching-learning process are supported by Digital Information and Communication Technologies (ICTs), such as forums, blogs, virtual learning environments, multimedia, slides, e-mail and e-books. Both are powered by the Internet (NOGUEIRA, 2011).Kenski (2007) explains the possibilities that technologies can offer in education:

> Technologies extend the possibilities of teaching beyond the short and limited space of physical presence of teachers and students in the same classroom. The possibility of interaction between teachers, students, objects and information involved in the teaching process redefines the whole dynamic of the classroom and creates new bonds between the participants (KENSKI, 2007, p. 88).

Educators, however, need to be open to this type of learning. The teacher's mentality needs to change so that he too can learn from the student and not be the holder of knowledge. It is understood that through interaction, students learn more and better (KENSKI, 2007).

According to Werneck (2002), change in teaching depends mainly on teachers, their training and the transformation of pedagogical practices in the classroom. In particular, the subject of Biology "can be one of the most relevant subjects and worthy of students' attention, or one of the most insignificant and unattractive subjects, depending on what is taught and how it is taught.". (KRASILCHIK, 2008, p. 15)

During their initial or continuing training, teachers need to understand the very process of constructing and producing school knowledge, understand the differences and similarities between the processes of producing scientific knowledge and school knowledge, know the characteristics of school culture, know the history of science and the history of teaching the science they work with and where they are related. These elements are just one of the characteristics of teaching work and, without ignoring the other dimensions, they already reveal and demonstrate its complexity (PEREIRA, 2006).

According to Moreira and Masini (1982), learning must be meaningful. This means that the teacher must take into account what the students already know, what they "bring from home" is of great value. In addition, David Paul

Ausubel's Meaningful Learning (19182008) believes in the influence of the environment on the subject. However, the authors emphasise that there is no point in developing a fun lesson if it is conducted automatically, without allowing for reflection and the negotiation of meanings. What is taught in the classroom must maintain a link with the student's life.

CHAPTER 3

METHODOLOGICAL FRAMEWORK

3.1 FOCUS

This study aims to analyse the environmental perception of Biological Sciences students and the teachers' knowledge of Environmental Education.

3.2 TYPE OF RESEARCH

The research design adopted in this scientific study is non-experimental, as the events investigated were observed without the intervention of the researcher. All the researcher had to do was collect the information, without any intervention from the variables.

The type of research was descriptive, as it describes the facts, phenomena and characteristics of a given population. This type of research used systemic data collection techniques such as questionnaires and interviews.

3.3 DELIMITATION

3.3.1 Temporal

Data was collected in the first and second semesters of 2013 and again in the second semester of 2016. In 2016, the students answered the same questionnaire applied in 2013.

3.3.2 Space

The space was restricted to the municipality of Caldas Novas, Goiás, Brazil, and specifically to the Biological Sciences course at the Faculdade de Caldas Novas - UNICALDAS.

3.4 POPULATION

All the students and teachers of the Biological Sciences course at the Faculdade de Caldas Novas-UNICALDAS.

3.5 SAMPLE

3.5.1 Inclusion criteria

Students who joined the course in the first and second semesters of 2013 and who had already studied Environmental Education - EE, in the 7th term, and Environmental Planning - PA, in the 8th term, will be included. Teachers from the Biological Sciences course will also be included in the sample.

3.5.2 Sample to be used

The 23 students enrolled in 2013 and these same students later in 2016, who had already studied Environmental Education (EE) in the 7th term and Environmental Planning (PA) in the 8th term. In addition to the students, 12 Biological Sciences lecturers from the second semester of 2016 took part in the research.

3.6 HYPOTHESIS

The hypothesis is that students entering the Biological Sciences course have an individualised and limited sensory environmental perception. It is assumed that during the course this notion of the environment and sustainability will be broadened and put into practice.

In addition, there is the assumption that the course, specifically the teachers, do not have the necessary knowledge and training on Environmental Education (EE). It is inferred that these teachers are not aware of the laws and documents that require environmental education in the classroom and that, as a result, they do not use efficient pedagogical practices.

3.7 DATA COLLECTION TECHNIQUES AND INSTRUMENTS

3.7.1 Interview

The interview, according to Gil (2008), is very suitable as a data collection technique for obtaining information about what people know, believe, expect, feel or want. An interview with pre-established questions was carried out with the teachers of the Biological Sciences course at the Faculdade de Caldas Novas - UNICALDAS, in order to ascertain their knowledge of Environmental Education and investigate the teaching practices used by these professionals.

3.7.2 Questionnaire

The questionnaire is a data collection instrument consisting of a series of ordered questions that must be answered in writing and without the presence of the interviewer (LAKATOS; MARCONI, 2003). A questionnaire was applied to students who joined the Biological Sciences course at the Faculdade de Caldas Novas - UNICALDAS in 2013. Later, in the last semester of 2016, the same questionnaire was used to analyse changes in environmental perception.

3.8 SOURCES OF INFORMATION

3.8.1 Primary Sources-Subjects

Interview applied to the teachers of the Biological Sciences course at the Faculdade de Caldas Novas - UNICALDAS, and the questionnaire applied to the students of this same course who entered in 2013.

3.8.2   Secondary Sources-Objects

Books, articles, monographs, theses, texts available on the Internet, databases, newspapers, magazines, films, among others that deal with environmental perception and environmental education, pedagogical practices, interdisciplinarity and meaningful learning.

3.9   HOW THE DATA COLLECTED WILL BE PROCESSED

The analysis of the data collected through the interviews with the teachers and the questionnaire applied to the students was represented both through graphs and in a descriptive manner, as it contained open and closed questions, in which the research participants could also leave their ideas and opinions in a discursive manner.

3.10   ETHICAL CONSIDERATIONS

It was not intended to use the interviewees' real names in order to preserve their identity. To this end, the last numbers of the enrolment were used to refer to the students *and the* term "Professor", followed by a Roman numeral to refer to the professors. Only the name of the Higher Education Institution and the course being researched will be used.

CHAPTER 4

ANALYTICAL FRAMEWORK

4.1 CHARACTERISATION OF THE RESEARCH SITE

This region is home to the largest hydrothermal spa in the world, so the main economic vocation is centred on tourism, which consequently drives hotels, construction, the property market, commerce and, more recently, the educational, cultural and, in an embryonic way, industrial sectors.

The research Higher Education Institution is called Faculdade de Caldas Novas - UNICALDAS and is located at Avenida Portal do Lago, Quadra 09, Lotes 01 to 28, Loteamento Portal do Lago, in the city of Caldas Novas - GO.

It offers private higher education courses, maintained by Sociedade de Ensino de Caldas Novas - LTDA, a legal entity governed by private law, registered with the CNPJ/MF under No. 3.494.082/0001-66, with its Articles of Association registered under No. 327 in book No. B-16, folios 014, at the 2nd Notary's Office in the city on 4 October 1999. It is governed by Report No. 23/2002 of the MEC's Higher Education Secretariat; Ministerial Order No. 646 of 7 March 2002, published in the DOU of 8/3/2002, Section 1, p. 25; Resolutions and Regulations issued by its Board of Governors, as well as documents drawn up by itself.

This HEI receives students from several neighbouring municipalities, including: Água Limpa, Caldas Novas, Corumbaíba, Ipameri, Morrinhos, Marzagão, Pires do Rio and Rio Quente. It runs undergraduate, postgraduate and extension courses. It aims to respond to social demands and seeks to exercise citizenship by valuing and respecting diversity and ethics, as well as contributing to the development of humanity.

According to the Course Pedagogical Project (PPC) (2016), UNICALDAS will offer the appropriate pedagogical development and practices so that Biology graduates achieve skills such as a generalist vision, being critical and ethical, aware of the need to act with quality and responsibility in favour of the conservation and management of biodiversity, health policies, the environment and biotechnology.

Its architectural structure is made up of 9 blocks and 60 classrooms, which accommodate six courses: Biology, Pedagogy, Environmental Engineering, Accountancy, Business Administration and Law. The teaching staff is made up of 68 professionals, 12 of whom are lecturers on the Biological Sciences course. There are currently 1,270 students, of which 45 are students on the course under study.

4.2 ANALYSING STUDENT QUESTIONNAIRES

In order to better visualise the sample of students on the Biological Sciences course at the Faculdade de Caldas Novas - UNICALDAS, it is worth pointing out that, as this is a private higher education institution, there is a considerable drop-out rate and "enrolment cancellations". There are also several students who "pay for subjects" due to the fact that the course doesn't have all the periods.

With this in mind, we organised a table (TABLE 2) showing the students who took part in this research. We took into account the students who had answered the questionnaire in 2013, and who had already studied Environmental Education - EE, in the 7th term, and Environmental Planning - PA, in the 8th term.

The students (1, 2, 4, 5, 12, 15, 19 and 23) highlighted in Table 2 are students who joined in the second half of 2013 and were placed in the first half of 2013 because it was a small class. This means that they were "fast-tracked". The questionnaire was also administered to them in 2013 and they had already taken the subjects set out in the previous paragraph.

Table 2 - Demonstration of the students who took part in the survey

N°	LAST LICENCE PLATE NUMBERS	SEX	ENTRY	ALREADY EA	ALREADY PA
T"	008	F	2013/2	YES	YES
2	032	F	2013/2	YES	YES
3	004	F	2013/1	YES	YES
4	003	F	2013/2	YES	YES
5	022	F	2013/2	YES	YES
6	014	F	2013/1	YES	YES
7	037	F	2013/1	YES	YES
8	017	M	2013/1	YES	YES
9	011	F	2013/1	YES	YES
10	001	F	2013/1	YES	YES
11	042	F	2013/1	YES	YES
12	005	F	2013/2	YES	YES
13	005	F	2013/1	YES	YES
14	007	F	2013/1	YES	YES
15	011	F	2013/2	YES	YES
16	026	M	2013/1	YES	YES
17	019	F	2013/1	YES	YES
18	040	F	2013/1	YES	YES
19	021	F	2013/2	YES	YES
20	044	F	2013/1	YES	YES
21	002	M	2013/1	YES	YES
22	016	M	2013/1	YES	YES
23	004	F	2013/2	YES	YES

Source: Organised by the author (2017).

The first question on the questionnaire sought to find out the gender of the students who entered in 2013. This can be seen in Table 2 above and Graph 1 on the following page:

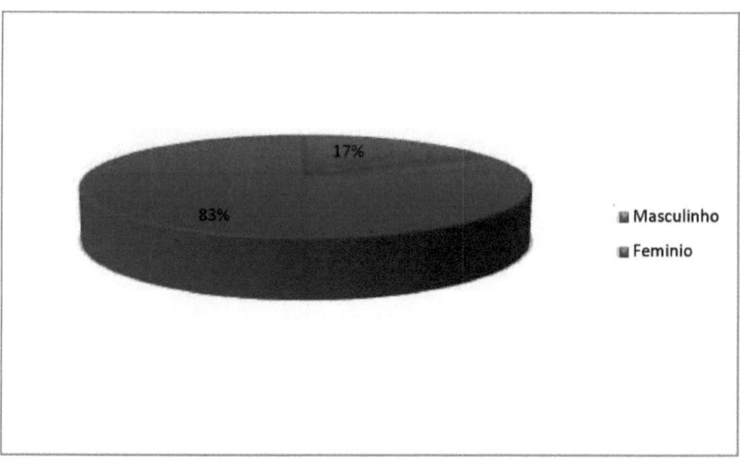

Graph 1 - Sex of the students who took part in the survey
Source: Organised by the author (2017).

Women are in the majority on the Biological Sciences course, both in the first semester of 2013 and in the second semester of the same year. 83 per cent of the incoming students in this period are female, while a further 17 per cent are male.

The second question in the questionnaire sought to find out the age of the students who joined the Biological Sciences programme in 2013. The result was that 39 per cent were under 18, 31 per cent between 19 and 25, 17 per cent between 26 and 30, 9 per cent between 31 and 40 and only 4 per cent over 40 (GRAPH 2).

It can be seen that the class surveyed is a young class, because if the two highest percentages are added together, it totals 70 per cent of the students with an average age of 20.

Garcia (2014), in the Correio Brasiliense newspaper, says that the Ministry of Education shows that the number of young people entering higher education courses at the age of 17 has increased by 60 per cent in four years. Scholars differ on the results of this advance, for some it is a positive point, but for others entering higher education too early is not good, because "It is necessary, before anything else, to have considerable maturity and life experience. This is what will guide the work and solidify the ethics of the future professional," points out Xênia Benfatti, a professor and researcher in the educational field at the University of Fortaleza (Unifor).

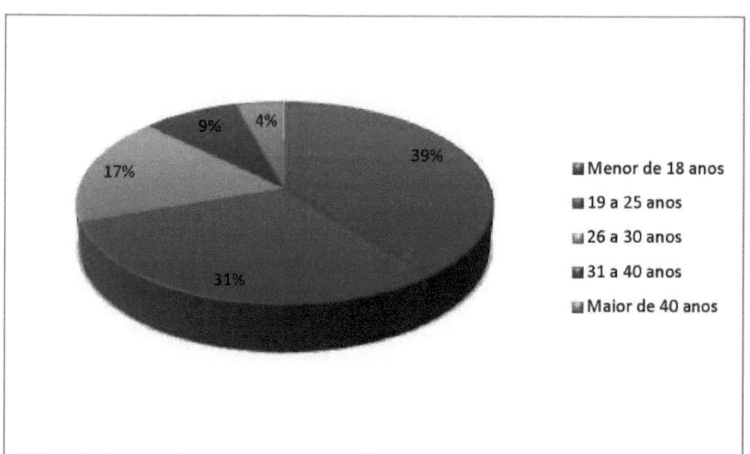

Graph 2 - Age of students when they joined the Biological Sciences programme
Source: Organised by the author (2017).

Question 3 asked the students if they had ever studied any other higher education course. None of them have another degree, only 03 students out of the 23 surveyed started another course (Law, Environmental Engineering and Pedagogy), at this same institution, and asked for a transfer because they didn't identify with the subjects.

Question 4 sought to find out what the "environment" is, hoping that the students would be able to understand that it involves a set of ecological units that function as a natural system, and include all vegetation, animals, microorganisms, soil, rocks, atmosphere and natural phenomena that can occur within the limits of human beings.

For this reason, the most appropriate answer would be letter "d) It is the place where living beings (plants, animals and human beings) live and relate to each other." Graph 3 parallels the answers given in 2013 and 2016. Only those most chosen by the students were used.

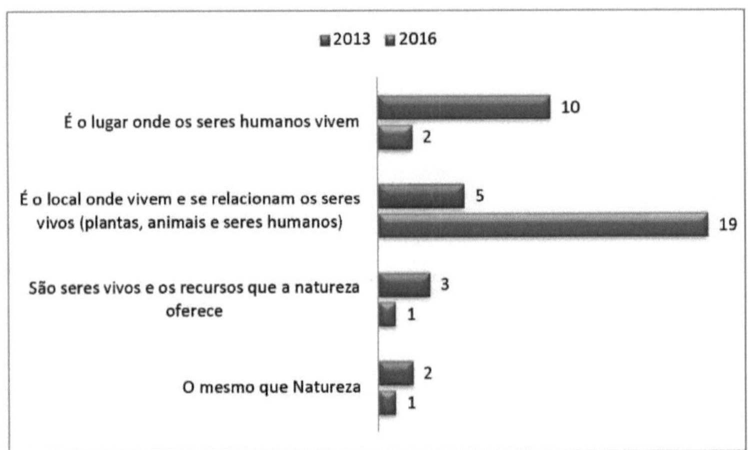

Graph 3 - Students' definition of the environment (2013 and 2016)

Source: Organised by the author (2017).

The concept of the environment was broadened significantly, because in 2013 the majority believed that it was restricted to the place where human beings lived, but in 2016 the majority chose to define it beyond the place where they live, adding the relationship between the beings that make it up.

Option "c) Animals and plants." is not shown in Graph 3, as it was not ticked by any of the students, either in 2013 or 2016.

In the fifth question, the students were asked to tick the alternative that showed elements that are part of the environment. The most appropriate option would be the letter "e) All the previous answers are correct". Graph 4 compares the answers given by the students in 2013 and later in 2016.

The option "d) hills, fields and its Faculty" does not appear in Graph 4, on the following page, as it was not ticked by any student, nor in

2013 nor in 2016. Only one student ticked the option "a) forest, rivers and his house".

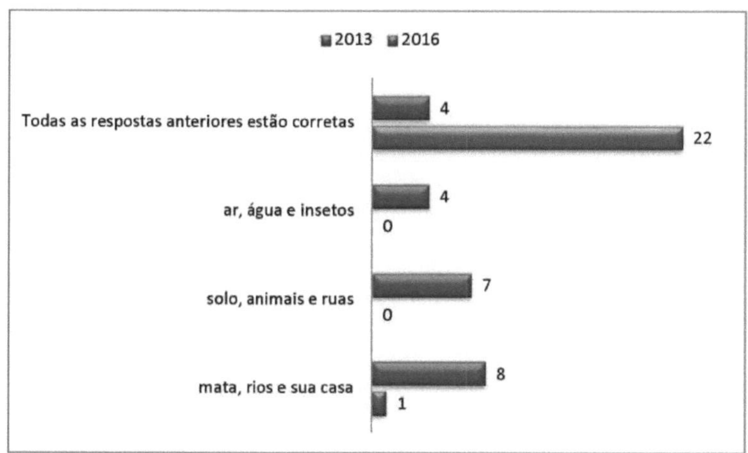

Graph 4 - Elements that make up the environment (2013 and 2016)

Source: Organised by the author (2017).

The students' perception of the elements that make up the environment expanded from 2013 to 2016. While in 2013 students chose only one of the questions that didn't fully answer the fair question, in 2016 the majority ticked the one that involved all the others listed above.

To see if the students were able to understand the water distribution process, they were asked the following question: "6) The water that reaches the taps of the residents of Caldas Novas comes from:".

The students were expected to tick "d) From the Pirapitinga River", information provided by the Municipal Water and Sewage Department (Demae) of Caldas Novas, Goiás state, and confirmed later in 2016.

Graph 5 compares the answers given by Biological Sciences students in 2013 and later in 2016.

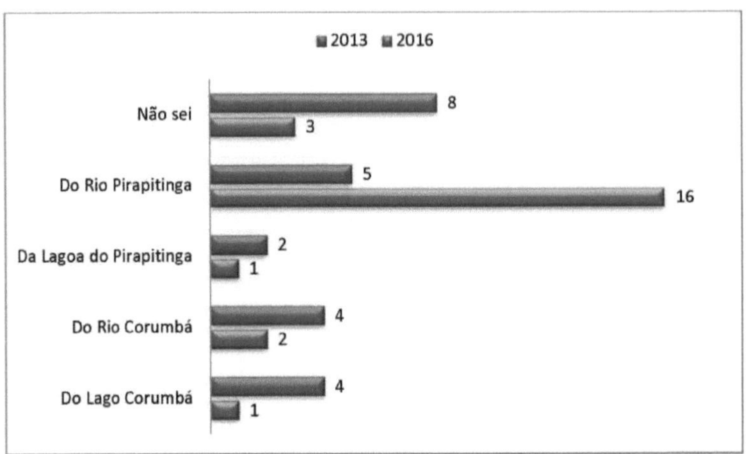

Graph 5 - Students' perception of the municipality's water supply (2013 and 2016)
Source: Organised by the author (2017).

In 2013, the number of students who recognised that they didn't know where the water that supplies the city came from was much higher than in 2016. Only 3 of the 23 students who answered the questionnaire in 2016 said they still didn't know the answer to this question.

The discrepancy between the correct answer, which would be the "Pirapitinga River", is clearly visible. In 2013, only 5 students ticked this option, while in 2016, 16 of the 23 surveyed did so.

There has been an evolution in the perception of students on the Biological Sciences course regarding the municipality's water supply. It's worth mentioning that in 2015, in the second semester, there was a field project for the "Health Programmes" subject, which gave the students a visit to the Caldas Novas Water and Sewage Treatment Plant, where the water supply process was explained.

The seventh question sought to ascertain the students' perception of soil. When defining the term "soil", option "d) Material derived from the decomposition of rocks by the action of physical or chemical agents, which may or may not contain organic matter" would be the most appropriate, as it is the most complete. The others also define it, but in a much more superficial way.

Graph 6 shows a parallel between the answers given by Biological Sciences students in 2013 and later in 2016.

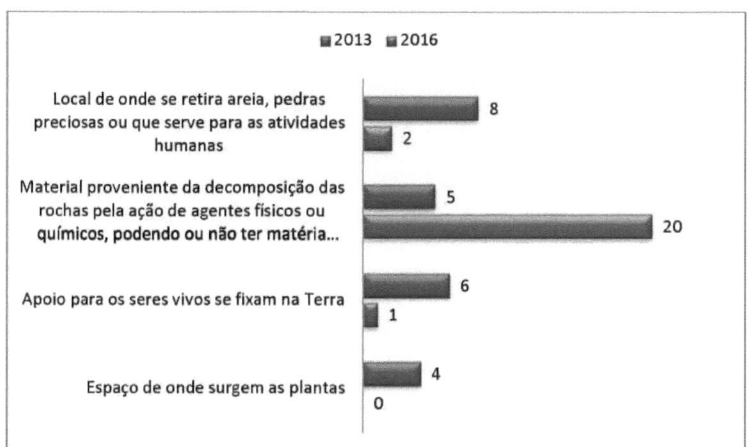

Graph 6 - Students' definition of soil (2013 and 2016)

Source: Organised by the author (2017).

In 2013, the majority of the students defined soil as a place where sand, precious stones or human activities take place. In 2016, 20 of the 23 students opted for the more complete and scientific definition.

Soil was perceived differently by the students surveyed between 2013 and 2016. It is inferred that this was due to the fact that they took Geology (5th period) and Palaeontology (6th period), both of which address issues related to the study of soil.

It should be noted that option "c) Region of the environment composed of mineral elements.". was not ticked by any of the students in any of the

of the years in which the questionnaire was administered. And the alternative "a) Space where plants come from" was chosen by 4 students in 2013, but none of them chose it in 2016.

The eighth question sought to understand the students' perception of air. To do this, it provided 5 options, the last of which was "e) All of the above answers are correct." the most complete alternative.

Graph 7 shows a comparison between the answers given by Biological Sciences students in 2013 and later in 2016.

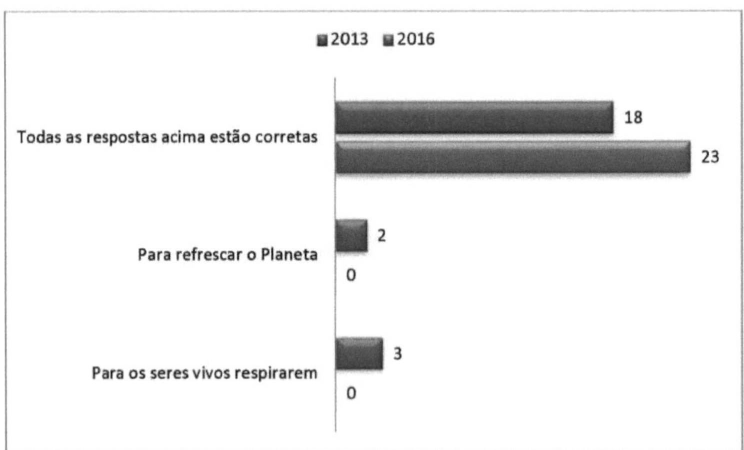

Graph 7 - Students' perception of air (2013 and 2016)
Source: Organised by the author (2017).

On this question, the discrepancy was smaller than on previous questions. In 2013, the majority of answers defined all the others as correct, and again in 2016. However, in 2016, all the students opted for the same answer as the appropriate one, showing that the five who chose other options in 2013 have changed their minds.

The ninth question sought to recognise the Biological Sciences students' perception of the term "forest". Graph 8 compares the answers given by the Biological Sciences students in 2013 and later in 2016.

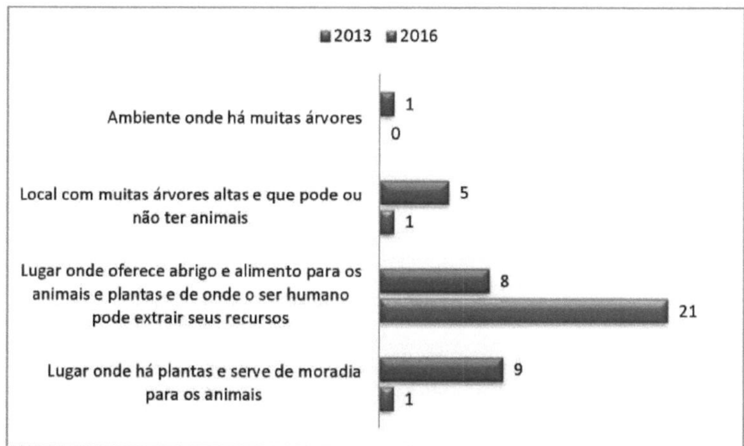

Graph 8 - Students' perception of the forest (2013 and 2016)

Source: Organised by the author (2017).

It's clear that the students on the Biological Sciences course have changed their perception of the definition of forest. They opted for the more straightforward answer in 2016.

It's worth pointing out that the forest is not part of the predominant vegetation in the Goiás region. There, the Cerrado is the predominant biome. There is, however, considerable growth in "planted forests" in Goiás. Although it is not personally recognised by most of the interviewees, it is covered by teachers, according to the syllabuses of the Ecology and Field Ecology subjects.

With this in mind, the tenth question asked: "What is the name of the predominant biome in the Caídas Novas region?". Students were expected to answer "b) Various types of Cerrado: cerradão; campestre;

gallery forest; rupestrian savannah", as this alternative is more complete than option "a) Cerrado".

Graph 9 compares the answers given by Biological Sciences students in 2013 and later in 2016.

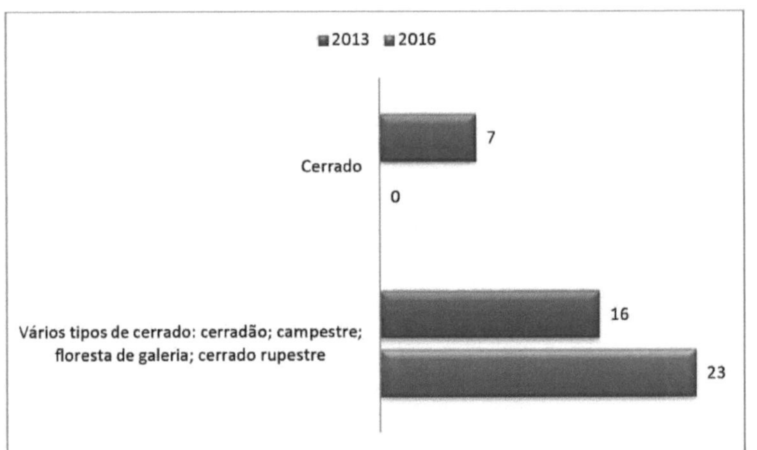

Graph 9 - Students' perception of the region's biome (2013 and 2008)
2016)
Source: Organised by the author (2017).

It was possible to see that the Biological Sciences students recognise that the Cerrado is the predominant biome in the Caldas Novas region. However, the difference was in recognising that this biome is made up of a variety of ecosystems and can be found in the form of: cerradão; campestre; floresta de galeria; cerrado rupestre and others.

Graph 9 shows that in 2013, seven students answered only "Cerrado" and in 2016, all of them were able to select the most appropriate, complete and scientific alternative. This topic is covered in several subjects, as it is part of the institution's objectives to prepare students to live and interact in society. These include the subject "Field Ecology I and II", offered in the 7th[o] and 8th[o] periods.

In question 11[a] of the questionnaire applied to the Biological Sciences students, the students were asked which phrase would best describe the Cerrado biome, which they ticked in the previous option.

The option that students were expected to choose was "b) Savannahs that present varied physiognomies, ranging from clean fields devoid of woody vegetation to dense tree formation", as the others extend this vegetation throughout the country.Graph 10 shows the answers given by Biological Sciences students in 2013 and later in 2016.

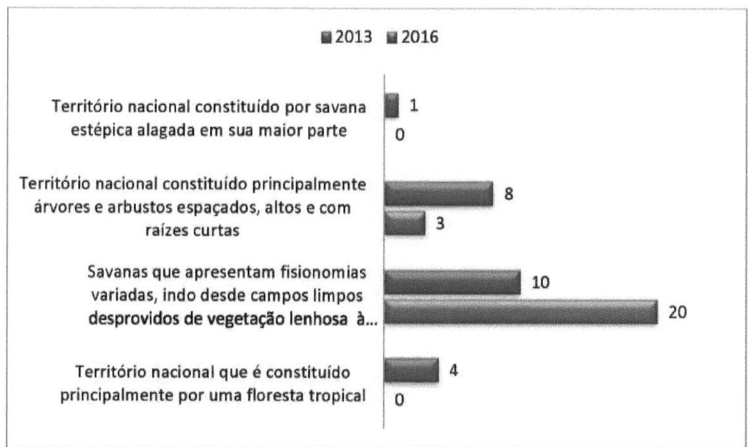

Graph 10 - Students' perception of the type of Cerrado vegetation
Source: Organised by the author (2017).

In 2013, most of the students answered that the Cerrado was a "national territory", but it is known that there are various types of biomes at national level, including forests, grasslands, steppes, woodlands and plant refuges, as well as the Cerrado, which covers 20 per cent of the Brazilian region.

In question 12[a] , students were asked to tick the option that described the most common animals in the Cerrado region. The option "a) jaguar, maned wolf, giant anteater, red deer" is the one that students were expected to tick, as the others have between 1 and 2 animals that are not part of this biome.

Graph 11 was drawn up by comparing the answers given by Biological Sciences students in 2013 and later in 2016.

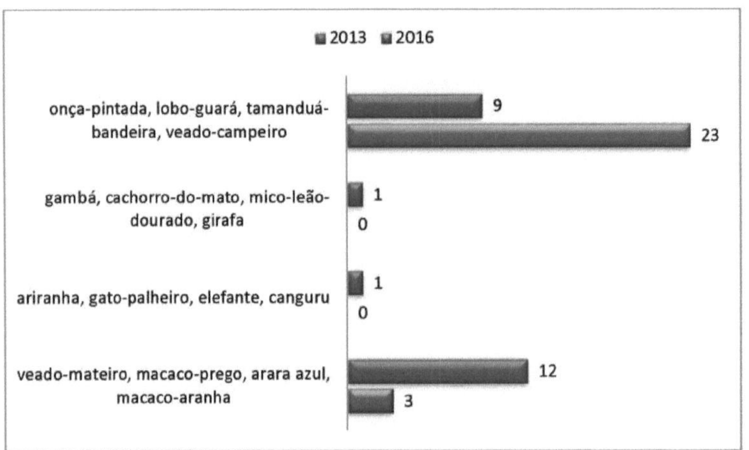

Graph 11 - Students' perception of Cerrado animals (2013 and 2016)
Source: Organised by the author (2017).

The second and third options in the graph above were only chosen in 2013. It is not correct, as it defines "giraffe", "kangaroo and elephant" as animals from the Cerrado. This assertion is incorrect and was not made in 2016.

In 2013, the students were divided between the first and last options in Graph 11. However, the "spider monkey" is an animal that inhabits the Amazon rainforest and the "roe deer" is found in humid forests in southern Brazil. Therefore, the correct alternative would be the first one shown in the graph. It can be seen that, in 2016, all the students marked it as the true option.

Question 13[a] asked Biological Sciences students the difference between rubbish and pollution. They were given five options to choose from, the most suitable of which was "e) None of the above is correct". The others weren't completely correct, as the definitions were very simple and unscientific.

Graph 12 compares the answers given by Biological Sciences students in 2013 and then in 2016.

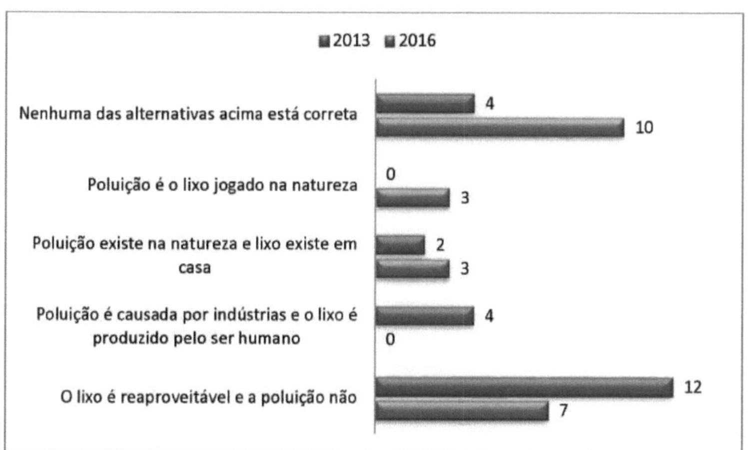

Graph 12 - Students' perception of rubbish and pollution (2013 and 2016)
Source: Organised by the author (2017).

As has already been said, none of the definitions of rubbish and pollution are correct. Even so, some students, even after studying biology, still confuse these terms.

According to Mano, Pacheco and Bonelli (2010), pollution is an ecological alteration caused by human beings that directly or indirectly harms their lives or well-being, causing damage to natural resources and impeding economic activities. Waste is any

waste arising from human activities or generated by nature in urban agglomerations.

Also according to Graph 12, there was an increase in the correct statement, from 4 to 10 students. The second option in the graph was not ticked in 2013, and in 2016 it was ticked by 3 students as being true. It can be seen that this subject still causes confusion among students.

Question 14[a] asked the students who was most responsible for environmental damage. It was hoped that the students would choose option "e) All of society", as this would cover all the others mentioned above.

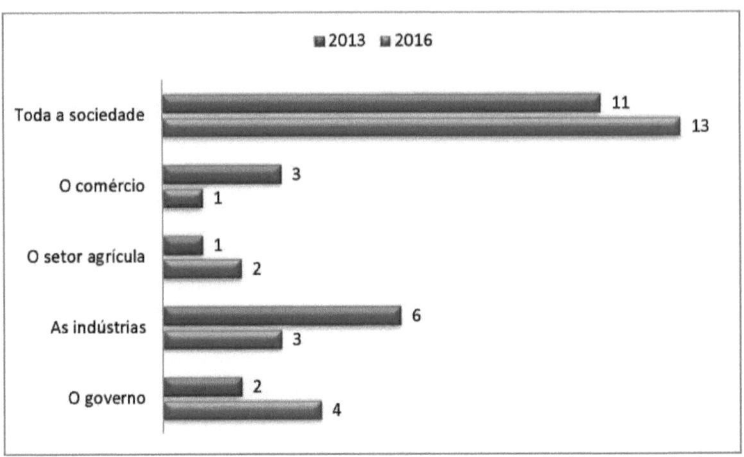

Graph 13 - Students' perception of those responsible for pollution (2013 and 2016)

Source: Organised by the author (2017).

Graph 13 above was organised to compare the answers given by Biological Sciences students in 2013 and later in 2016.

The majority of students chose the correct option in both 2013 and 2016. The "agricultural sector" and the "government" received more votes in the last questionnaire than in the first. Bearing in mind that there are several types of pollution (air, water, soil, etc.) it would be inappropriate to delegate responsibility to just one of the sectors in the question.

Question 15[a] asked students what they would do with the rubbish they produce at home. This question doesn't have a correct option, but "b) Separate what is recyclable" would be the most ecologically correct, along with composting.

Graph 14 organises and compares the answers given by Biological Sciences students in 2013 and later in 2016.

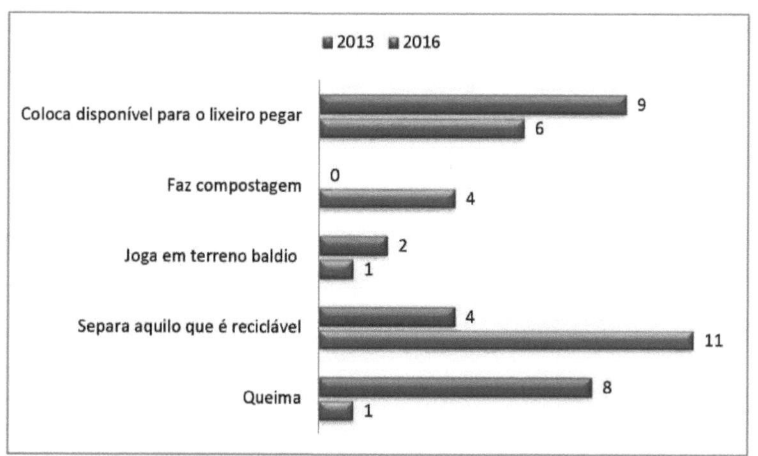

Graph 14 - Destination of rubbish produced by students (2013 and 2016)

Source: Organised by the author (2017).

There has been a significant increase in the amount of rubbish that is sorted at home and sent for recycling. There has also been a reduction in rubbish that was made available for regular rubbish collection. The amount of rubbish that was burnt fell from 8 to 1.

In this question, space was made available for justification and some of the students wrote: "I burn paper and rubbish from the toilet"; "I throw building waste on waste ground and when I cut down a tree"; "I put it in the toilet"; "I put it in the rubbish bin".

only what I can't reuse at home"; "I collect the recyclables and a girl comes round to my house every fortnight to pick them up".

In 2013, none of the students ticked the "composts" box, which is probably a term they weren't aware of yet. In 2016, 4 students ticked this option and one of them explained: "My composting is simple, done on the ground in my backyard, but I use it regularly."

Question 16ª asked Biology students what they do when they change a mobile phone, computer, notebook or tablet that no longer works. The most appropriate answer would be "d) Return it to the shop I bought it from".

Graph 15 analyses the answers given by Biological Sciences students in 2013 and then in 2016.

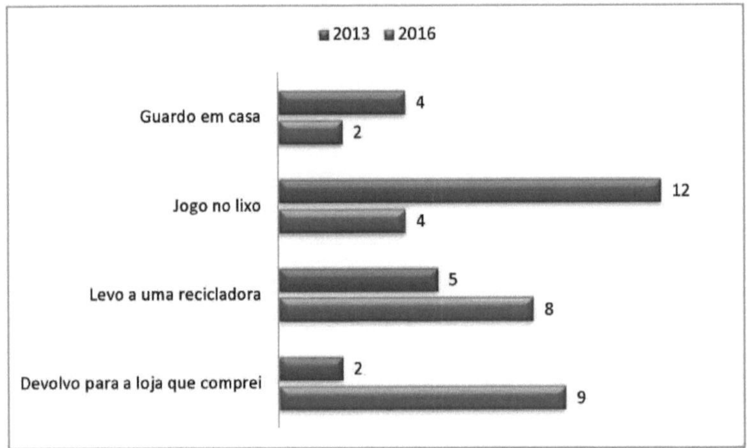

Graph 15 - Destination of electronic devices by students (2013 and 2016)
Source: Organised by the author (2017).

The question is relevant because according to the United Nations Environment Programme (UNEP), of the 118 chemical elements in the periodic table, mobile phones contain 43 of them, such as mercury, cadmium and

lead, which are toxic metals. For this reason, they should not be disposed of in the same way as household rubbish.

Graph 15 shows that there has been a change in students' behaviour when it comes to disposing of electronic devices. In 2013, most assumed that they were thrown in the rubbish bin, while in 2016 they reported returning them to the shop they bought them from or taking them to a recycling centre.

In question 17[a] the students were asked if they had ever read any books on Environmental Education. The aim was to find out what knowledge they had about Environmental Education when they started the course, and what had changed when they left. It is understood that the practice of reading, from this perspective, enables the development of the subject's critical thinking, making them more critical, emancipated and autonomous.Graph 16 was drawn up to analyse the answers given by the Biological Sciences students in 2013 and later in 2016.

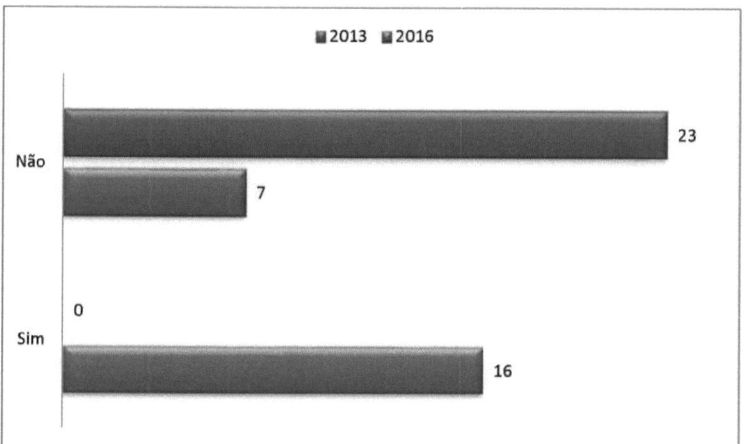

Graph 16 - Students' reading on Environmental Education (2013 and 2016)
Source: Organised by the author (2017).

The results of this graph show that in 2013 the students had not read any books on Environmental Education, but in 2016,

Sixteen students said they had had contact with this type of literature. This shows that there has been an increase in their knowledge.

Question 18ª asked teachers which professional would be best suited to teach environmental education in schools. In order to analyse the answers given by Biological Sciences students in 2013 and later in 2016, Graph 17 below was put together.

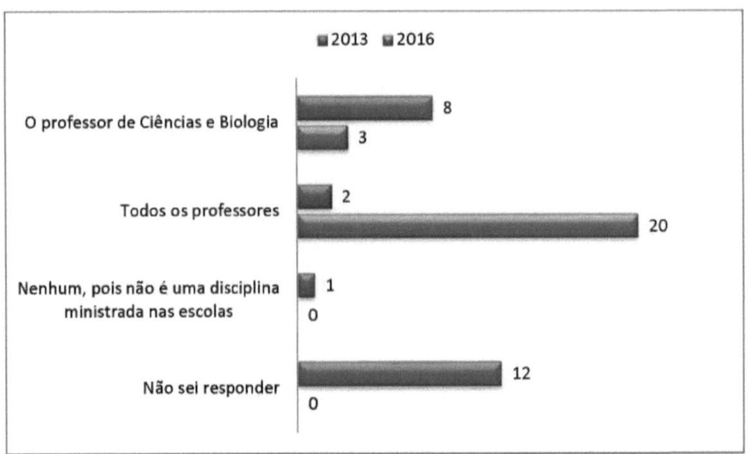

Graph 17 - Who would be responsible for Environmental Education in the students' view (2013 and 2016)
Source: Organised by the author (2017).

It can be seen in the graph above that in 2013, when they entered the Biological Sciences course, the students didn't know who was responsible for Environmental Education at school, or even believed that it was only the Biology teacher's responsibility.

However, in 2016, the students' perception of Environmental Education in the school context changed, as 20 students said that this would be the collective responsibility of all the teachers. Only three of them replied that it was the Biology teacher's responsibility.

Question 19[a] asked the students if they sorted their rubbish at home. According to the Ministry of the Environment, the separation of household rubbish relieves landfill sites and landfill sites of the waste that can't be reused, so much of the solid waste generated at home can be reused.

The question therefore becomes relevant, since the practice of recycling saves natural resources and generates income for rubbish collectors, part of the population that depends on discarded solid waste to survive.

Graph 18 analyses the answers given by Biological Sciences students in 2013 and later in 2016.

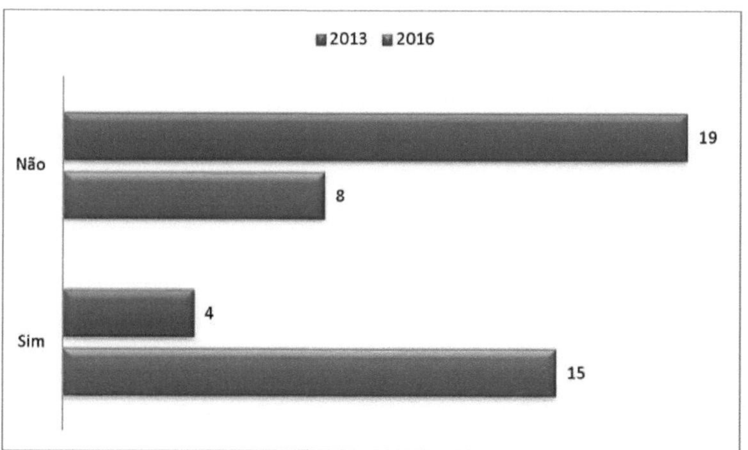

Graph 18 - Students who separate household waste (2013 and 2016)

Source: Organised by the author (2017).

In 2013, recycling was not a priority for the students. This can be seen by analysing the fact that 19 students answered that they did not separate their household waste. In 2016, this rate increased from 8 students who answered "Yes" in the first questionnaire to 15 who gave the same answer in the second.

It is worth emphasising that it was not possible to ascertain whether these assertions were true. However, even if the students don't really separate the rubbish, as they said in the survey, it is possible to see that they are now able to realise the importance of this act.In question 20[a] , the aim was to find out from the Biological Sciences students what they think is needed to improve the population's quality of life.The question is discursive and so there were several answers, but the most commonly used in the two years were scored. Graph 19 analyses the answers given by Biological Sciences students in 2013 and later in 2016.

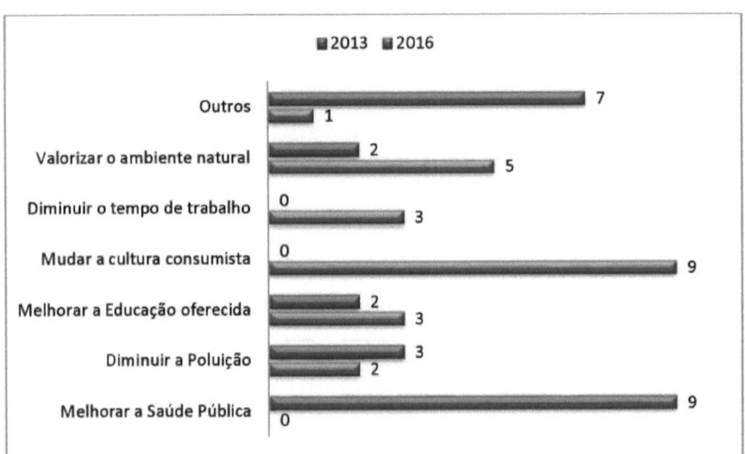

Graph 19 - Actions that can improve the population's quality of life (2013 and 2016)
Source: Organised by the author (2017).

This question is relevant to this study because it allows us to see whether students have a global and complete view of society and how it is made up. It is understood that merely identifying problems is easier than proposing solutions, since the latter requires much greater scientific and theoretical knowledge.

It can be seen that the opinion of these students has changed over the course of the course. In 2013, they believed in more concrete actions on the part of the population, such as "reducing pollution" and "improving public health" and "improving education", which involved the public authorities. However, at the end of their studies, this concern became more individualised, such as "reducing consumerism" and "working hours".

Question 21[a] of the questionnaire applied to the students asked them how they assessed the teaching practices used by their teachers. It's worth noting that even though they were in the first term of the course, this question was counted.

Graph 20 shows the answers given by Biological Sciences students in 2013 and later in 2016.

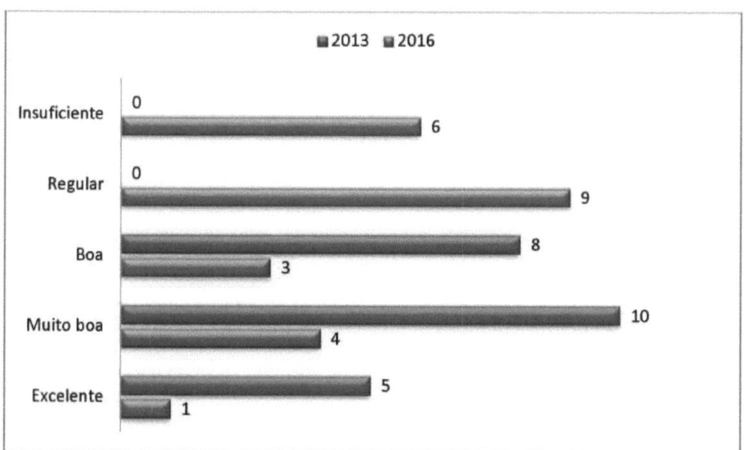

Graph 20 - Evaluation of teachers' pedagogical practices from the students' point of view (2013 and 2016)
Source: Organised by the author (2017).

Pedagogical practices, as already discussed in the theoretical framework of this work, are of the utmost importance in order to awaken meaningful learning in the student, and through this, changes in their behaviour can occur.

thoughts, attitudes and behaviour. This question therefore becomes relevant, as students are the best people to assess the methodology and didactics used in the classroom.

Graph 20 shows that at the start of the course, students rated their teachers' pedagogical practice as "Very Good" and "Good", while at the end of the course this perception changed to "Fair" and "Insufficient".

It is inferred that this discrepancy is due to the fact that students only become aware of what "pedagogical practice" really is after the Interdisciplinary Practices courses offered in the 4th, 5th, 6th and 8th periods.

The last question, number 22, asked the students how they rated the classes in Environmental Education and Environmental Planning (7th and 8th periods). The answers are shown in Graph 21 below.

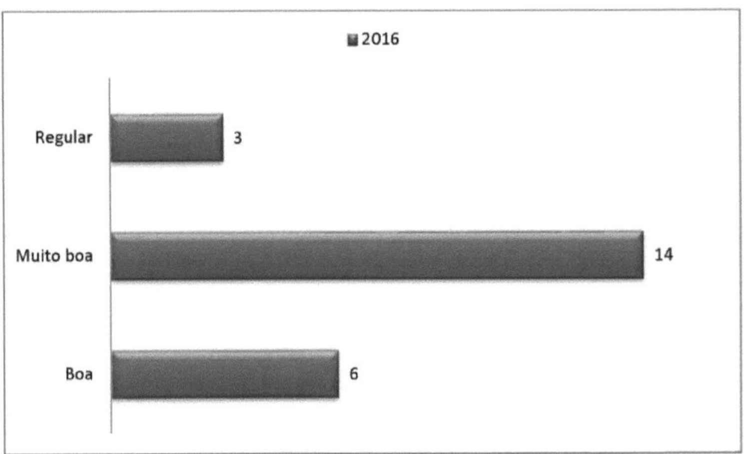

Graph 21 - Students' assessment of the subjects Environmental Education and Environmental Planning (2016)
Source: Organised by the author (2017).

It's worth remembering that this question was added to the initial questionnaire because the subjects are only studied in the 7th and 8th periods of the course, making it impossible to answer in 2013.

This question is relevant because it is through the subjects of Environmental Education and Environmental Planning that students will have direct contact with the environment and the educational actions that can protect it. Therefore, knowing how these classes are perceived by students is of paramount importance for re-evaluating educational practice and the approaches taken in them.

Graph 21 shows that the majority of students (14) say that these classes can be considered "Very Good". Among the justifications for this assessment, we can cite some left by the students:

> "The teacher uses documentaries and this makes us aware of the emergence of this science."
> "I find it interesting that the teacher is able to concretise everything she explains in theory."
> "Field classes make us put into practice everything we study in the classroom."
> "I rate it Very Good because he doesn't just stick to the blackboard or books."
> "It was through these lessons that we realised the importance of teaching about the environment in schools."
> "Good, but it could be improved, I think we could have more classes where

The students who answered "Fair" justified this by saying that there is a lack of motivation during lessons and that teachers should offer more interdisciplinary work. One of these students suggested a project as a solution for motivation and interdisciplinarity in these classes.

4.3 ANALYSING THE INTERVIEWS WITH TEACHERS

In 2016/2, all the lecturers on the Biological Sciences course at the Caldas Novas College in Rio Grande do Sul were interviewed. So 12 teachers were interviewed. Of this percentage, 4 are men and 8 are women. The majority have worked at Unicaldas for more than five years. The sample is around 40 years old and all specialise in an area of education or the sciences. There are 3 Masters in the course: 1 in Ecotone Ecology, 1 in Science and 1 in Linguistics, the latter being the only one with a PhD in the same area. The number of Masters and Doctors would be higher if the institution recognised training outside the country. The teachers at the Caldas Novas College - UNICALDAS were asked if they had studied Environmental Education in their degree courses. Only those with degrees in the sciences answered "yes". Even though Environmental Education is in fact an interdisciplinary subject, it is restricted to courses that deal with the environment.

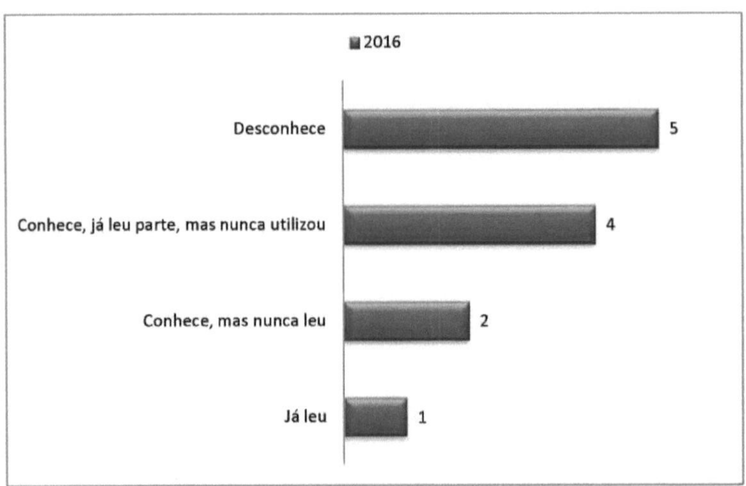

Graph 22 - Teachers who know the Cross-Cutting Themes (2016)
Source: Organised by the author (2017).

He asked the teachers if they had ever taken a specific course on the environment, sustainability or environmental education. Of the 12, only one gave a positive answer. The teachers were asked if they were familiar with the Cross-Cutting Themes and if they had ever read and/or used them (GRAPH 22). The same question was asked in relation to the Curriculum Parameters and the results are shown in Graph 23 below.

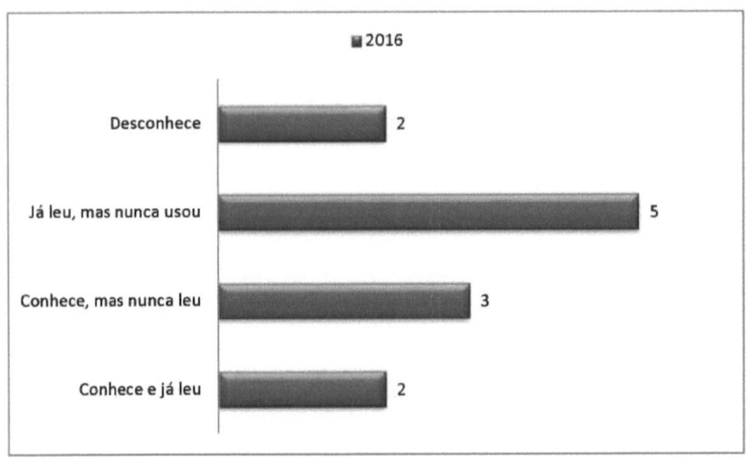

Graph 23 - Teachers who know the Curriculum Parameters (2016)
Source: Organised by the author (2017).

The previous questions are highly relevant as they assess the knowledge of

Biological Sciences teachers in relation to the documents that underpin primary and secondary education. It is understood that in order to offer efficient Environmental Education, it is necessary to know what is established in the documents and guidelines that permeate this teaching.

It can be seen in the previous graphs (GRAPHS 22 and 23) that the PCNs are better known than the Cross-Cutting Themes. It is possible to see that teachers have "heard of it", but that the majority have never read or used any of these documents.

It should be emphasised that because the Biological Sciences course is offered at degree level, where professionals are prepared to teach Science and Biology classes in both primary and secondary schools, the higher education teachers involved in preparing these professionals must be familiar with the skills that this future educator will develop.

The teachers were asked if they had ever taken any real events into the classroom to discuss with their students.

This question was asked in order to ascertain the pedagogical practices used by teachers at the HEI investigated. Problematisation is widely used at all levels of education, as an interdisciplinary practice, in which the "facilitator", in this case the teacher, offers a problem, or brings up a subject that is in vogue, to be discussed with their students, making a parallel with their subject.

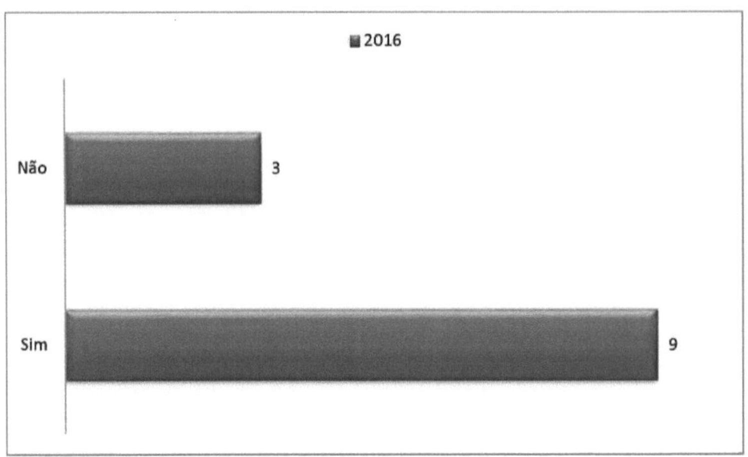

Graph 24 - Teachers who use the problem technique as a teaching practice (2016)

Source: Organised by the author (2017).

Most of the teachers replied that they had already used some kind of problem to teach their classes. When asked how this happened, the answers were: "Through newspaper reports", "Magazine reports", "Television interviews", "Problems that occurred in their own municipality".

The teachers interviewed were also asked which pedagogical practices they use most to teach their classes. This question is organised in Graph 25.

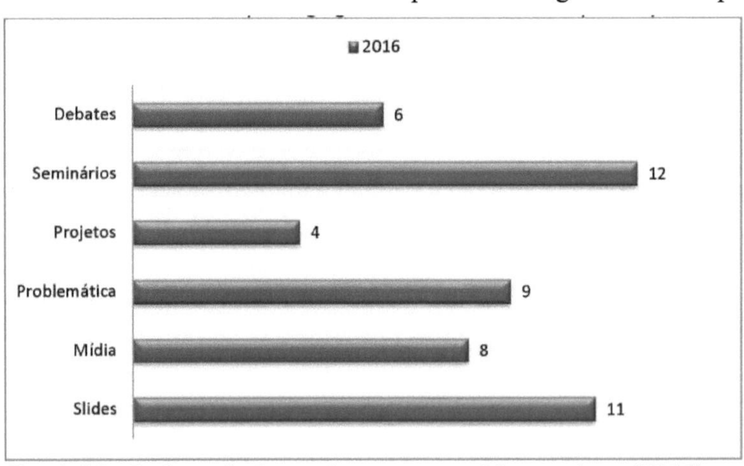

Graph 25 - Pedagogical practices most used by teachers (2016)

Source: Organised by the author (2017).

It was found that the practices most used by Biological Sciences teachers are the use of slides and the promotion of seminars. Next, starting the lesson by raising a problem, a media, debates and, lastly, the use of projects.

It was not possible to ascertain whether the practices reported by these professionals actually take place. However, if we compare the answers given by the students in Graph 20, in which the students rate these practices as "regular" and "insufficient", we can see that there is a lack of consistency.

It sought to find out from the course teachers whether they believed that the subject of Environmental Education was sufficient to bring about changes in the behaviour of students (undergraduates), both in their daily lives and when teaching in primary and secondary schools.

The majority replied that this subject "alone" or in isolation cannot bring about changes in student behaviour. The teachers believe that the combination of all the subjects can "yes" bring about changes in students' attitudes. The teachers were asked how they evaluated their environmental attitude. The answers are organised in Graph 26 below.

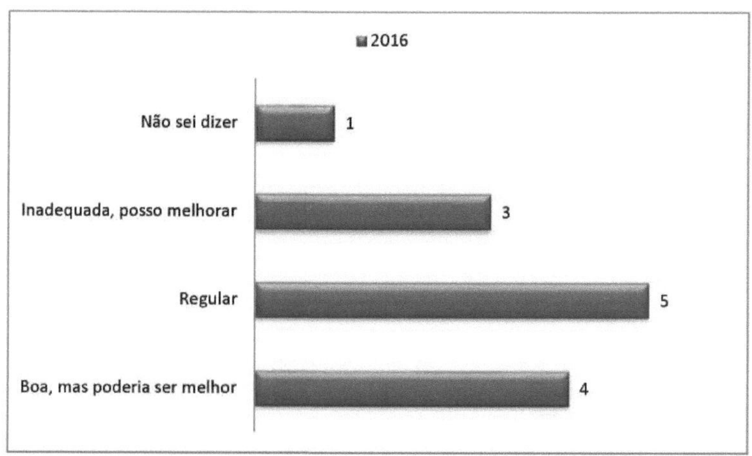

Graph 26 - Teachers' self-assessment of their environmental attitude
Source: Organised by the author (2017).

It's worth noting that during the interview, some examples of environmental attitudes were given, such as: "recycling rubbish; restoring degraded forests; protecting the source of a river; not throwing rubbish in the wrong places; using public transport more; reducing the use of plastic bags, etc.".

During the interview, the teachers rated their environmental attitude as "fair" and both recognised that they should and could change.

With this in mind, we sought to find out whether the Caldas Novas College - UNICALDAS has any specific projects on the environment that take place on a regular basis. The relevance of this question is due to the fact that as a Higher Education Institution, according to the theoretical framework of this study, it should maintain interdisciplinary work, research, extension and/or intervention projects that provide undergraduates with the opportunity to broaden their knowledge and put into practice what they learn in the classroom, in theory.

CHAPTER 5

CONCLUSION

5.1 FINAL CONSIDERATIONS AND RECOMMENDATIONS

Perception is the ability to apprehend something through the senses or the mind. Human beings acquire a perception through the influences of the environment in which they live. Environmental perception is no different; it studies how man perceives the environment in which he lives and how he interacts with it.

The study of Environmental Perception emerged alongside movements to care for and preserve the environment. Scholars and defenders of nature, including philosophers, noticed that each person, despite living in the same environment as the other, understood the environment differently. With this in mind, the need for environmental education was realised.

The 1960s to 1980s were marked by strong impacts on man's relationship with nature, which sounded alarm bells, drawing the world's attention to the exhaustion of natural resources and energy sources, jeopardising life in society. It was at the International Conference on the Human Environment in Stockholm/Sweden in 1972 that the Charter of Principles on the Human Environment was drawn up, which included environmental education among its 26 articles.

Since then, the concepts of Environmental Education have been disseminated and expanded considerably. It was introduced into schools through the National Curriculum Parameters (1997) and complemented with the Transversal Themes. The aim is to change the culture of a people who have lived for hundreds of years thinking that natural resources are inexhaustible. The aim is to awaken in people, from an early age, the care and zeal that must be taken to maintain a balance between nature and man.

In higher education, environmental education plays a complementary role to primary and secondary education. It is understood that when students arrive at university, they already have an ecological vision of the environment in which they live. This is where practices are inserted - intervention projects, research, extension projects and actions of direct intervention in the environment and the community - so that students can put into practice the knowledge they have acquired throughout the teaching and learning process.

The Biological Sciences course provides an understanding of the emergence of life and its organisation over time, under the action of evolutionary processes. What's more, when it comes to a degree programme, it seeks to train professionals who are qualified to teach Science and Biology, and it is inferred that these future teachers will be able to provide appropriate Environmental Education for their students.

It is therefore understood that the training of higher education teachers must be sufficient to bring about significant changes in the learning and behaviour of these undergraduates. This is so that when they reach the labour market (classroom) they can teach something that they have already experienced and put into practice. With this in mind, the culture of preserving and caring for the planet will be disseminated and practised.

When investigating Environmental Education in Higher Education, it was possible to see that the matrix or content offered alone is not enough to bring about an environmental transformation in the subject. It was found that the pedagogical practices used by teachers have a considerable influence on undergraduates' learning. Actions such as an "interdisciplinary project" are more motivating and make learning more enjoyable. Others such as "field classes" give students the opportunity to put into practice what they have learnt in theory.

During the research carried out with the students of the Biological Sciences course at the Faculdade de Caldas Novas-UNICALDAS, it was possible to ascertain that there has been an evolution in their environmental perception, from the moment they joined the course in 2013 to its finalisation in 2016.

The perception of caring for the environment - surrounding and natural - has changed; attitudes towards preservation and conservation are more noticeable in 2016 than in 2013; scientific and theoretical knowledge related to the environmental area, added to practice, has been able to provoke significant changes in students' thinking and behaviour.

However, it was found that the students on this course are not entirely satisfied with the pedagogical practices used by higher education lecturers. In 2013, these different methods used by teachers were rated "Good" or "Very Good", while in 2016 they were rated "Regular" and "Insufficient". It is inferred that these teachers should reflect on this issue.

BIBLIOGRAPHICAL REFERENCES

ARAÚJO, Monica Lopes Folena; OLIVEIRA, Maria Marly de; NOGUEIRA, Romildo Albuquerque. **The Pedagogical Practice in Environmental Education of Biology Teachers in Porto Velho - RO**. 2005. Available at: <http://www.nutes.ufrj.br/abrapec/vienpec/CR2/p2.pdf>. Accessed on: 14 January 2017.

BARBIERE, J. C.; SILVA, D. Sustainable Development and Environmental Education: a common path with many challenges. **Revistas de adminstração Marckenzie**. São Paulo, v. 12, n. 3, p. 51-82, May/June 2011. ISSN 1678-6971.

BERNA, V. **How to do environmental education**. São Paulo: Paulus, 2004.

BRAGA, Maria Lúcia Santaella. **Perception**: a semiotic theory. 3. ed. São Paulo: Brasiliense, 1998.

BRAZIL. Constitution (1988). **Constitution of the Federative Republic of Brazil**. Organised by Alexandre de Moraes. 16. ed. São Paulo: Atlas, 2000.

. **CONAMA RESOLUTION No. 1** of 23 January 1986. Published in the DOU of 17 February 1986, Section 1, pages 2548-2549.

_______. Law No. 9.795, of 27 April 1999. **National Environmental Education Policy** (DF): Senado Federal; 1999.

_______. **National Curriculum Parameters**: environment. Department of Basic Education. Brasília, DF, 1997.

_______. **Cross-cutting themes**. National Curriculum Parameters. Secretariat for Primary Education. Brasília: MEC/SEF, 2001.

_______. Ministério da Educação (MEC), Secretaria de Educação Média e Tecnológica (Semtec). **PCN Ensino Médio**: orientações educacionais complementares aos Parâmetros Curriculares Nacionais - Ciências da Natureza, Matemática e suas Tecnologias, Brasília: MEC/Semtec, 2002.

______. **Curriculum Guidelines for Secondary Education** (OCEM): Natural Sciences, Mathematics and their Technologies. 2006. Available at: <http://portal.mec.gov.br/seb/arquivos/pdf/book_volume_02_internet.pdf>. Access

______. **Mapping Environmental Education in Brazilian Higher Education Institutions:** elements for public policies. Technical Documents Series, No. 12. Management Body of the National Environmental Education Policy. Brasília-DF. 2007.

CABRAL, Fabiana Ferreira et al. Environmental perception of 6th grade public school students. REMOA - **Revista Monografias Ambientais** - Universidade Federal de Santa Maria, v. 14, n. 2, mai-ago. 2015, p.151-161.

BRAIN. The world inside your head. **Brain magazine.** 2010. Available at: <http://cerebro.weebly.com/uploads/3/2/2/3/3223841/revista.pdf>. Accessed on: 20 October 2016.

COIMBRA, José de Ávila Aguiar. Language and environmental perception. In: PHILIPPI-JR, Arlindo; ROMERO, Marcelo de Andrade; BRUNA, Gilda Collet (Orgs). **Environmental Management Course.** Barueri: Manole, 2004, p. 525-570.

DEL RIO, Vicente; OLIVEIRA, Lívia. **Environmental perception:** the Brazilian experience. 2. ed. São Paulo: Studio Nobel, 1999.

DIAS, G. F. **Educação ambiental:** princípios e praticas. 6. ed. São Paulo: Gaia, 2000.

DORNFELD, Carolina Buso. Environmental Education: reflections and challenges in Higher Education. 2015. Available at : <http://www.foar.unesp.br/Home/projetoviverbem/educacao-ambiental-reflexoes-e- desafios-no-ensino-superior---resumo.pdf>. Accessed on: 14 January 2017.

FERNANDES, Joice; REZENDE FILHO, Cyro de Barros. **Environmental Perception.** São Paulo: Editora CRV, 2010.

FERNANDES, R. S., PELISSARI, V. B., et al. Environmental perception of university students. **Preservation Magazine:** The Environment in Espírito Santo. Year I, No. 2, Dec 2002 to February 2003.

FERREIRA, C. R. T. **Evaluation of urban environmental degradation through environmental perception:** The case of the upper Limoeiro basin, Presidente Prudente, SP. Master's dissertation. Postgraduate programme in Geosciences. University of Presidente Prudente. SP. 2001.

FERREIRA. Aurélio Buarque de Holanda de. **Dictionary of the Portuguese Language.** Editora Positivo-Livros, edition 1/2014).

FORTUNATO, I.; SHIGUNOV NETO, A. S. **Environmental education and teacher training.** Rio de Janeiro: Editora Ciência Moderna, 2015.

FRAZÃO, Juliana Oliveira; SILVA, Jobson Martins da Silva; CASTRO, Carla Soraia Soares de. Environmental Perception of Students and Teachers in the Preservation of Sea Turtles at Pipa Beach - RN. **Electronic Journal Master in Environmental Education.** ISSN 1517-1256, v. 24, January to July 2010.

FRÓIS, Kátia P. Review of the perception of architecture in Bruno Zevi, Christian Norberg-Schulz and through Heidegger to the phenomenological possibility of Merleau-Ponty. **Revista OLAM Ciência**

& **Tecnologia**, Rio Claro, v. 1, n. 2, p. 123-152, nov. 2001.

GADOTTI, M. **Pedagogy of the Earth.** 4. ed. São Paulo: Petrópolis, 2000.

GIL, A. C. **Como elaborar projetos de pesquisa.** 4. ed. São Paulo: Atlas, 2008.

GUIMARÃES, M. **A formação de educadores ambientais.** São Paulo: Papirus, 2005.

HENRIQUES, Ricardo et al. (eds.) Environmental education: learners of sustainability. **SECAD Notebooks1**. Ministry of Education - Secretariat for Continuing Education, Literacy and Diversity, Brasilia: March 2007.

HERNÁNDEZ, F. **Transgression and change in education:** work projects. Porto Alegre: ArtMed, 1998.

HOCHBERG, Julian E. **Perception.** Translated by Álvaro Cabral. Rio de Janeiro: Zahar, 1973.

KENSKI, V. M. **Educação e Tecnologias:** o novo ritmo da informação - Campinas, SP: PAPIRUS, 2007.

KRASILCHIK, M. **Práticas do ensino de Biologia.** 4. ed. rev. and expanded. SP - Editora da universidade de São Paulo. 2008.

LAKATOS, E. M.; MARCONI, M. A. **Fundamentos de metodologia científica.** 5. ed. São Paulo: Atlas, 2003.

LEE, Terence. **Psychology and the environment.** Translated by Álvaro Cabral. Rio de Janeiro: Zah ar, 1977.

MANO, Eloisa Biasotto; PACHECO, Élen Beatriz A. V.; BONELLI, Cláudia Maria Chagas. **Environment, Pollution and Recycling.** 2 ed. São Paulo: Bucher, 2010.

MARIN, Andreia Aparecida. **Research in Environmental Education.** Pesquisa em educação ambiental e percepção ambiental. vol. 3, n . 1 - pp. 203- 222, 2008.

MELO, Vera L. M. O. Landscape from the perspective of new geographical approaches. In: **Meeting of Latin American Geographers**, 10, 2005, São Paulo. Proceedings. São Paulo: University of São Paulo, 2005. p.9146-9165.

MONTEIRO, Manuela Matos; FERREIRA, Pedro Teixeira. **Being human.** Porto/Portugal: Porto Editora, 2009

MOREIRA, M. A.; MASINI, E. A. F. **Significant learning:** David Ausubel's theory. São Paulo: Editora Moraes, 1982.

NICOLESCU, Basarab. **The manifesto of transdisciplinarity.** São Paulo: TRIOM, 1999.

NOGUEIRA, Luciana de Medeiros. **The influence of new technologies in the school context.** Annals of SILEL. Volume 2, Number 2. Uberlândia: EDUFU, 2011.

OKAMOTO, Jun. **Environmental Perception and Behaviour.** São Paulo: Plêiade, 1996.

OLIVEIRA, Lívia de. Perception of the environment and geography. **Revista OLAM Ciência e**

Tecnologia, Rio Claro, v. 1, n. 2, p. 14-28, nov. 2001.

PASCARELLI. N.F. **Educando para Preservação da Vida.** São Paulo: Wak Editora, 2011.

PAZIN, F. A.; SCARPELINI, S. **Simulation:** definition. Ver Medicina. 2007; 40(2):162-166.

PEREIRA, J. E. D. **Formação de professores:** pesquisas, representações e poder. 2. ed. Belo Horizonte: Autêntica, 2006.

PERLS, Frederick; RALPH, Hefferline; GOODMAN, Paul. **Gestalt therapy.** Summus Editorial. 1997.

REIGOTA, M. **Meio ambiente e representação social.** 7. ed. São Paulo: Cortez, 2007.

GARCIA, Daniela. **University students getting earlier and earlier.** Correio Brasiliense. 2014. Available from:<http://www.correiobraziliense.com.br/app/noticia/eu-estudante/ensino_ensinosuperior/2014/08/03/ensino_ensinosuperior_interna,440531/univ ersitarios-cada-vez-mais-precoces.shtml>. Accessed on: 15 Feb. 2017.

GERALDINO, C. F. G. **A definition of the environment.** GEOUSP - Espaço e Tempo (Online), São Paulo, v. 18, n. 2, p. 403-415, 2014. Available at: <http://www.revistas.usp.br/geousp/article/view/84540/87454>. Accessed on: 14 January 2017.

GUNTZEL-RISSATO, Cíntia et al. (Orgs). **Environmental Education and Public Policies:** Concepts, Foundations and Experiences. São Paulo: Appris, 2013.

SAAD, F. D. **Demonstrations in science:** Exploring air and liquid pressure phenomena through simple experiments. São Paulo: Editora Livraria da Física, 2005.

SATO, Michèle. **Environmental education.** São Carlos: RIMA, 2003.

SIBBEL, A. Pathways towards sustainability through higher education. **International Journal of Sustainability in Higher Education**, v. 10, n. 1, p. 68-82, 2009.

SILVA, C. W. M.; LYRA, L. H.; ALMEIDA-CORTEZ, A. S. "Educação ambiental contribuindo para a preservação da mata de Dois Irmãos, Recife-PERevista **Mestrado em Educação Ambiental**, Rio Grande, v. 15, 2003.

SOARES, Edvaldo. **Scientific methodology:** logic, epistemology and norms. São Paulo: Atlas, 2003.

SOBRAL, Eliane da Silva. **Environmental perception of university students in Brazil and Portugal.** Master's dissertation, Nove de Julho University - UNINOVE, 2014.

SPIRONELLO, R. L.; TAVARES, F. S. ; SILVA, E. P. Environmental education: from theory to practice, in search of sensitisation and environmental awareness. **Revista Geonorte**, v.3, n.4, p.140-152, 2012.

TRISTÃO, M. The dimensions and challenges of environmental education in the knowledge society. In: RUSCHEINSKY, A. (Org). **Environmental education:** multiple approaches. Porto Alegre: ArtMed, 2002.

UNESCO. **United Nations Decade of Education for Sustainable Development.** Final Document: International Plan of Implementation. Brazil: UNESCO, 2005, p.113.

WERNECK, Hamilton. **We teach too much, we learn too little.** Editora Vozes, Petrópolis 2002.

WHITE, R. T. **Conceptual and conceptional change.** Learning and Instruction, Oxford, v. 4, p. 117-121, 1994.

ZABALA, Antoni. **Globalising approach and complex thinking:** a proposal for the school curriculum. Porto Alegre: ARTMED, 2002.

APPENDIX A - QUESTIONNAIRE FOR STUDENTS ON THE BIOLOGICAL SCIENCES COURSE

1) Sex: (   ) F (   ) M

2) Age: ____

3) Have you taken any higher education courses? __________________

4) What is the environment?
   a) (   ) The same as Nature.
   b) (   ) They are living beings and the resources that nature offers.
   c) (   ) Animals and plants.
   d) (   ) It's the place where living beings (plants) live and interact, animals and human beings).
   e) (   ) It's the place where human beings live.

5) Which of these alternatives shows elements that are part of the environment?
   a) (   ) woods, rivers and his house.
   b) (   ) soil, animals and streets.
   c) (   ) air, water and insects.
   d) (   ) hills, fields and its Faculty.
   e) (   ) All the previous answers are correct.

6) The water that reaches the taps of Caldas Novas residents comes from the city's water supply:
   a) (   ) Lake Corumbá.
   b) (   ) The Corumbá River.
   c) (   ) The Pirapitinga Lagoon.
   d) (   ) The Pirapitinga River.
   e) (   ) I don't know.

7) What is soil?
   a) ( ) Space from which plants grow.
   b) ( ) Support for living beings to settle on Earth.
   c) ( ) Region of the environment composed of mineral elements.

d) (   ) Material from the decomposition of rocks by the action of
physical or chemical agents, and may or may not contain organic matter.
e) ( ) A place from which sand or precious stones are extracted or which is
used for human activities.

8) Why air?
a) (   ) For living beings to breathe.

b) (   ) As a source for energy production.
c) (   ) To refresh the planet.
d) (   ) To help plants reproduce.
e) (   ) All of the above answers are correct.

9) What is a forest?
a) ( ) A place where there are plants and where animals live.
b) ( ) An environment where there are many trees.
c) ( ) A place that provides shelter and food for animals and plants and from
which humans can extract their resources.
d) ( ) A place with many tall trees that may or may not have animals.

10) What is the name of the predominant biome in the Caldas Novas region?
a) (   ) Cerrado.
b) ( ) Various types of savannah: cerradão; grassland; gallery forest;
rupestrian savannah.
c) (   ) Atlantic Rainforest.
d) (   ) Caatinga.
e) (   ) Pampa.
f) (   ) Pantanal.
g) (   ) Coastal marine.

11) Which of the following phrases describes this biome?
a) ( ) National territory that is mainly made up of tropical rainforest.
b) ( ) Savannahs with a variety of physiognomies, ranging from clear fields
devoid of woody vegetation to dense tree formations.
c) ( ) National territory, consisting mainly of forest along the coast.
d) ( ) National territory made up mainly of spaced-out, tall trees and shrubs
with short roots.
e) ( ) National territory made up of mostly flooded steppe savannah.

12) Which of the options below only shows animals from this biome?
a) ( ) Jaguar, maned wolf, giant anteater, roe deer.
b) ( ) Skunk, bush dog, golden lion tamarin, giraffe.
c) ( ) giant otter, pussycat, elephant, kangaroo.
d) ( ) Fallow deer, capuchin monkey, hyacinth macaw, spider monkey.

13) What's the difference between rubbish and pollution?
   a) ( ) Pollution is rubbish thrown into nature.
   b) ( ) Pollution exists in nature and rubbish exists at home.
   c) (   ) Pollution is caused by industries and rubbish is produced by humans.
      human.

   d) ( ) Waste is reusable and pollution is not.
   e) ( ) None of the above is correct.

14) Who is most responsible for damaging the environment?
   a) () The government
   b) (   ) Industries
   c) () The agricultural sector
   d) (   ) Trade
   e) (   ) All of society

15) Where does your household rubbish go? (You can tick more than one
option, as long as you justify it below).
   a) (   ) Burning
   b) (   ) Separate what is recyclable
   c) (   ) Throw it on waste ground
   d) (   ) Compost
   e) (   ) Put it out for the binman to pick up
Justification: ____

16) What do you do when you change a mobile phone, computer, notebook or
tablet that no longer works?
   a) (   ) I keep it at home.
   b) (   ) I throw it in the bin.
   c) (   ) I take it to a recycling centre.
   d) (   ) I return it to the shop I bought it from.
   e) (   ) Other:____

17) Have you read any books or texts on environmental education?
(   ) Yes   () No (   ) Other: _______________________________

18) Who do you think is responsible for teaching environmental education in
schools?
   a) (   ) The science and biology teacher.
   b) (   ) All the teachers.
   c) (   ) None, because it's not a subject taught in schools.
   d) (   ) I can't answer that.

19) Do you separate rubbish at home? How do you do it?
(   ) Yes   () No (   ) Other:_______________________________

Justification: ____

20) What do you think is needed to improve people's quality of life?

21) How do you rate the teaching practices used by your teachers? Start your sentence with one of the words (Excellent, Very Good, Good, Fair, Insufficient) and justify your answer.

22) How would you rate the classes in Environmental Education (7th period) and Environmental Planning (8th period)? Start your sentence with one of the words (Excellent, Very Good, Good, Fair, Insufficient) and justify your answer.

APPENDIX B - INTERVIEWS WITH BIOLOGICAL SCIENCES TEACHERS

1) Sex: (   ) F (   ) M

2) Age: ___________________

3) Academic background: _______________________________________

4) Length of service at the institution and in the course: _______________

5) Subject you teach:____

6) Were there any specific subjects on the environment in your degree programme?
(   ) Yes   () No (   ) Other: _________________________________

7) Have you ever taken a specific course on the environment, sustainability or environmental education?
(   ) Yes   () No (   ) Other: _________________________________

8) Are you familiar with the Cross-Cutting Themes? Have you read them? Have you used it?

9) Are you familiar with the National Curriculum Parameters on Science and the Environment? Have you read them?

10) Have you ever taken a real-life event into the classroom to discuss with your students? Why?

11) What pedagogical practices do you use to teach your classes and how do

you evaluate them?

12) Do you believe that the subject of Environmental Education is sufficient to bring about changes in students' behaviour, both in their daily lives and when teaching in primary and secondary schools?

13) How do you rate your attitude towards the environment?

14) Does Faculdade de Caldas Novas have any specific projects on the environment that take place on a regular basis?

Printed by Books on Demand GmbH, Norderstedt / Germany